Teacher's guide to Book 1

GW00630749

CAMBRIDGE
UNIVERSITY PRESS

PUBLISHED BY THE PRESS SYNDICATE OF THE UNIVERSITY OF CAMBRIDGE
The Pitt Building, Trumpington Street, Cambridge, United Kingdom

CAMBRIDGE UNIVERSITY PRESS
The Edinburgh Building, Cambridge CB2 2RU, UK
40 West 20th Street, New York, NY 10011–4211, USA
10 Stamford Road, Oakleigh, Melbourne 3166, Australia
Ruiz de Alarcón 13, 28014 Madrid, Spain

http://www.cup.cam.ac.uk
http://www.cup.org

Printed in the United Kingdom at the University Press, Cambridge
Typeface Minion   *System* QuarkXPress®
A catalogue record for this book is available from the British Library

ISBN 0 521 77794 1 paperback

Illustrations by Robert Calow and Steve Lach at Eikon Illustration and Jeff Edwards
Cover image by Eikon Illustration
Cover design by Angela Ashton

Contents

The following people contributed to the writing of the SMP Interact
key stage 3 materials.

Ben Alldred	Ian Edney	John Ling	Susan Shilton
Juliette Baldwin	Steve Feller	Carole Martin	Caroline Starkey
Simon Baxter	Rose Flower	Peter Moody	Liz Stewart
Gill Beeney	John Gardiner	Lorna Mulhern	Pam Turner
Roger Beeney	Bob Hartman	Mary Pardoe	Biff Vernon
Roger Bentote	Spencer Instone	Peter Ransom	Jo Waddingham
Sue Briggs	Liz Jackson	Paul Scruton	Nigel Webb
David Cassell	Pamela Leon	Richard Sharpe	Heather West

Others, too numerous to mention individually, gave valuable advice,
particularly by commenting on and trialling draft materials.

Editorial team:	David Cassell	Project support:	Melanie Bull
	Spencer Instone		Tiffany Passmore
	John Ling		Martin Smith
	Mary Pardoe		Ann White
	Paul Scruton		
	Susan Shilton		

Introduction

What is distinctive about *SMP Interact*?

SMP Interact sets out to help teachers use a variety of teaching approaches in order to stimulate pupils and foster their understanding and enjoyment of mathematics.

A central place is given to discussion and other interactive work. Through discussion with the whole class you can find out about pupils' prior understanding when beginning a topic, can check on their progress and can draw ideas together as work comes to an end. Working interactively on some topics in small groups gives pupils, including the less confident, a chance to clarify and justify their own ideas and to build on, or raise objections to, suggestions put forward by others.

Questions that promote effective discussion and activities well suited to group work occur throughout the material.

SMP Interact has benefited from extensive and successful trialling in a variety of schools. The practical suggestions contained in the teacher's guides are based on teachers' experiences, often expressed in their own words.

Which pupils is it suitable for?

The material is written for pupils who are likely to achieve national curriculum levels 3 upwards at the end of key stage 3.

How are different levels of attainment catered for?

Three tiers are identified in the material:

- ▼ T(riangle): national curriculum levels 3 to 5 in key stage 3
- ■ S(quare): national curriculum levels up to 6 in key stage 3
- ● C(ircle): national curriculum levels up to 7/8 in stage 3

Books 1 and *N* contain all three tiers; this is so that they can be used in mixed attainment or setted classes. The symbols are shown like this:

   ▼□○ T only   ▼■○ T and S but not C, and so on

Pupils working mainly on T material will often be able to tackle some of the S material. Similarly, pupils working mainly on S material will often be able to use C material.

After *Books 1* and *N* there is a separate series for each tier, with overlapping material.

How are the skills of using and applying mathematics developed?

There are problems and investigations throughout the material, linked to the topics being studied. In addition there are pieces of work focusing specifically on skills such as working systematically, writing up results, and so on.

How are the pupils' books intended to be used?

The pupils' books are a resource which can and should be used flexibly. They are not for pupils to work through individually at their own pace. Many of the activities are designed for class or group discussion.

Activities intended to be led by the teacher are shown by a solid strip at the edge of the pupil's page, and a corresponding strip in the margin of the teacher's guide, where they are fully described.

Some teacher-led sections are intended for only one or two of the three tiers ▼ (T), ■ (S) or ● (C). In a mixed ability class you may need to group pupils on the basis of attainment when using these sections.

A broken strip at the edge of the page shows an activity or question in the pupil's book that is likely to need teacher intervention and support.

Where the writers have a particular way of working in mind, this is stated (for example, 'for two or more people').

Where there is no indication otherwise, the material is suitable for pupils working on their own.

Starred questions (for example, *C7) are more challenging.

What use is made of software?

Points at which software (on a computer or a graphic calculator) can be used to provide effective support for the work are indicated by these symbols, referring to a spreadsheet, graph plotter or dynamic geometry package respectively. Other suggestions for software support can be found on the SMP's website (see back cover).

How is the attainment of pupils assessed?

The interactive class sessions provide much feedback to the teacher about pupils' levels of understanding.

Each unit of work begins with a statement of the key learning objectives and finishes with questions for self-assessment ('What progress have you made?'). The latter can be incorporated into a running record of progress.

Longer term retention is assessed by tests covering all the work up to and including a particular unit. Revision questions similar to the test questions are included in periodic reviews in the pupil's book.

What will pupils do for homework?

Practice questions (on sheets with a 'P' before their reference number for *Book 1* and *Book N* and in booklets thereafter) can be used for homework. Some schools have bound together copies of practice sheets, sometimes with material of their own, into termly homework books.

Often a homework can consist of preparatory or follow-up work to an activity in the main pupil's book.

Using this teacher's guide

There are notes for each unit, following this pattern:

- an overview table to help you with your planning (except for the smaller units)
- a panel listing essential and optional items
- notes on the sections including preparation (in a panel) and points of guidance (each main point beginning with a ◊)

For reasons of economy, where pupils have to write their responses on a resource sheet the answers are not always shown in this guide. For convenience in marking you could put the correct responses on a spare copy of each sheet and add it to a file for future use.

General guidance on teaching approaches

Getting everyone involved

When you are introducing a new idea or extending an already familiar topic, it is important to get as many pupils as possible actively engaged.

Posing key questions
A powerful technique for achieving this is to pose one or two key questions, perhaps in the form of a novel problem to be solved. Ask pupils, working in pairs or small groups, to think about the question and try to come up with an answer. For example, before beginning work on place value and ordering decimals, ask pupils to decide whether each of the following statements is true or false:

$1.30 = 1.3$

1.38 is greater than 1.8

$1.3 \times 10 = 1.30$

When everyone has had time to work seriously at the problem (have a further question ready for the faster ones), you can then ask for answers, without at this stage revealing whether they are right or wrong (so that pupils have to keep thinking!). You could ask pupils to comment on each other's answers.

Open tasks
Open tasks and questions are often good for getting pupils to think, or think more deeply. For example, 'Working in groups of three or four, make up three questions for which multiplication is needed and three for which division is needed. Try to make your questions as varied as you can.'

Questioning skills

Sensitive questioning can help pupils develop their understanding.

Questioning with the whole class
Your own expectations will affect pupils' responses. If you ask closed questions and always greet the correct answer with 'That's right – well done', then pupils who were not right will feel they have nothing to contribute. Instead you can ask a pupil how they got their answer, without saying whether it is right or wrong. Then ask another pupil. In this way both correct and incorrect answers can help everyone learn.

Prompting
You may need to prompt a pupil who finds it difficult to respond to a question. Prompts might consist of

- encouragement – 'Just have a guess; you can have another go if it's wrong.'

- rephrasing – perhaps simplifying the wording

- asking a simpler question – if a pupil is stuck trying to divide 8 by 4, ask 'What do you get when you divide 8 by 2 – how many 2s in 8?'

Probing more deeply It is often important to probe more deeply, even when a pupil has answered apparently correctly. For example, if the question was 'I had £2.50 and spent £1.15. How much did I have left?' and you ask a pupil how they worked out the answer, they might say 'I subtracted.' Ask how they subtracted. One may have used a paper and pencil method. Another may say 'It's 85p to make £1.15 up to £2, and then there's another 50p to add.' Without further probing you would have missed valuable information about the pupils' methods.

Working in groups

Before organising group work, you may need to consider the following points.

Preparing In group work especially, pupils need to know where to get necessary resources. Then you are less likely to be interrupted constantly. Pupils also need to be clear about the purpose of the task so that they can work in a focused way.

Sometimes you may want one pupil from each group to report back to the class. It is usually a good idea if that pupil knows who they are at the start.

You may need to rearrange the furniture, so your class will need to be able to do this when you ask them.

Organising the groups Different activities may require different groups. For example, sometimes you may want pupils of similar attainment to work together; at other times you may want to mix attainments or allow pupils to choose their own groupings.

The group size is important. Groups of more than four or five can lead to some pupils making little or no contribution.

For some activities, you may want pupils to work unassisted. But for many, your own contribution will be vital. Then it is generally more effective if, once you are sure that every group has got started, you work intensively with each group in turn.

After the group work One way to help pupils feel that they have worked effectively is to get them to report their findings to the whole class. This may be done in a number of different ways. One pupil from each group could report back. Or you could question each group in turn. Or each group could make a poster showing their results.

Managing discussion

Discussion, whether in a whole-class or group setting, has a vital role to play in developing pupils' understanding. It is most fruitful in an atmosphere where pupils know their contributions are valued and are not always judged in terms of immediate correctness. It needs careful management for it to be effective and teachers are often worried that it will get out of hand. Here are a few common worries, and ways of dealing with them.

What if …   '… the group is not used to discussion?'

- Allow time for pupils to work first on the problem individually or in small groups, then they will all have ideas to contribute.

'… everyone tries to talk at once?'

- Set clear rules. For example, pupils raise their hands and you write their name on the board before they can speak.

'… a few pupils dominate whole-class discussion?'

- Precede any class discussion with small-group discussion and nominate the pupils who will feed back to the class.

- Set a limit on the number of times an individual can speak.

'… one pupil reaches the end point of a discussion immediately?'

- Tell them that the rest of the group need to be convinced and ask the pupil to convince the rest of the group.

- Accept the suggestion and ask the rest of the group to comment on it.

Starters (S1 to S12)

This is a collection of activities suitable for use in the first two or three weeks of year 7. Their purposes are

- to give pupils of all abilities an enjoyable and confident start to mathematics in the secondary school
- to give you a chance to get to know the pupils and how they work
- to help establish classroom routines and ways of working (whole class, group, individual)
- to give opportunities for homework

They do not not need a high level of number skill, so should be widely accessible.

There will not be time to do all the Starters at the beginning of year 7. Some may be left for later, perhaps as starting points for related units of work.

However, you may wish to consider including these in the initial selection, for the reasons given:

S5 'Finding your way'
   The topic is not addressed anywhere else in the material.

S3 'Two-piece tangrams' or S12 'Shapes on a dotty square'
   These use shape language and involve presentation skills.

S8 'Patterns from a hexagon'
   This gives early practice in using drawing instruments.

S6 'Gridlock'
   This is a good way of finding out about basic arithmetic skills.

Spot the mistake (p 4)

T

These two resource sheets give pupils of all abilities an opportunity, in a light-hearted context, to spot some mathematical errors – and some non-mathematical ones! It is more fun for pupils to work in pairs, rather than on their own.

Essential

Sheets 45, 46

---

1 Off to Benidorm in June

Fish tank framework is an impossible object.

Table left-hand rear leg is longer than the others.

Vase on the table is an impossible object.

Sofa is an impossible object.

Mirror reflection is incorrect; 'TAXI' and the clock face are the wrong way round.

Front door has handle and hinges on the same side.

The '33' above the front door is the wrong way round.

Vacuum cleaner plug has only one pin.

Vacuum cleaner hose has an extra hose tangled in it.

Triangles on shelf: right-hand one is an impossible object.

View of window blind is impossible.

2 Taxi to the airport

Clock has a back-to-front 3, and 8 where it should be 9.

P (parking) sign is on a road with double yellow lines.

Stop sign on road – S is wrong way round.

The word STOP (and the road marking) is on the wrong side of the road.

Bicycle has no front wheel.

Airport sign says 5 cm (centimetres).

Traffic lights have a right turn arrow to a no-entry street.

Rollerblader has one ice skate on.

Right-hand no-entry sign is pointing upwards.

Pedestrian crossing markings on the road should be rectangles.

Pedestrian crossing beacon is the wrong shape.

Street lamp on top of the no-entry sign is facing up.

A vegetarian butcher would not do much business!

Low bridge sign says min(imum) and should say max(imum).

Taxi is going to the airport, so should have turned left.

('Tax to rise 150%' is not necessarily wrong – it could!)

3 At the airport

A plane is flying upside down.

The plane taking off has no tail wings.

Passengers are walking along the wing of the waiting plane.

Waiting plane has RAF insignia on tail.

Waiting plane has a ski instead of a wheel.

Wind socks are blowing in opposite directions.

Tannoy message says 'train' instead of 'plane'.

Tannoy message contradicts time on clock.

Christmas tree contradicts Easter eggs sign.

Tax free sign: you cannot save 200%.

Suitcases are ticketed to Rome, and flight is to Benidorm.

Suitcase on weight machine has a square wheel.

Luggage weight sign says WAIT, not WEIGHT.

Luggage weight is in g(rams) and should be kg (kilograms).

The 'All departures' sign points to a no entry corridor.

You cannot 'Ski the Pyramids'.

You cannot 'Ice skate the Amazon'.

4 The hotel reception

Clock reads 14:60, and the minutes must be less than 60.

Sign above toilet doors says 'Welcome to BeMidorN.'

Calendar on reception desk says 31 June (only 30 days in June).

A Christmas tree in June is rare!

Toilet door pictures are the wrong way round.

Right-hand toilet door has handle and hinges on the same side.

Change sign – 100 pts should be 1000 pts.

Double rooms are cheaper than single rooms.

Atlantic views should read Mediterranean views in Benidorm.

Left-hand rear leg on table is longer than other three.

Tickets on luggage have changed since the airport to Roma and Home.

Plant doesn't sit in its flower pot.

Lift is on ground floor (the lowest in the list above the door) but shows on the list (and the call buttons) as going further down.

List of floors above lift door is missing floor 4.

Sign in lift mirror – S is wrong in reflection.

Sign in lift has a weight limit that is silly.

(The vase on the table is not an impossible object!)

5 By the pool

You don't get whales in the Mediterranean.

The flag at the top of the boat's mast should be blowing forwards.

Speed boat and water skier are not connected.

Plane is flying backwards if it is pulling the banner.

Banner should read 24 hours not 26.

Weather vane NSEW are wrong.

If the time is 23:30 the sun would not be out.

Temperature of $^-26^\circ$C is a bit chilly for sunbathing.

Water level in man's jug should be horizontal.

Sun lounger nearest to front is missing a leg.

Gazebo on the left-hand corner of the balcony is an impossible object.

Diving board heights are in millimetres, and should be in metres.

Lower diving board is at a greater height (8) than the upper (4).

There appears no means of access to the lower diving board.

'Do not feed fish' sign is unlikely in a swimming pool.

Man fishing not possible (we hope) in a swimming pool.

Swimming pool depth signs are wrong.

Shark in swimming pool.

Stairs and railings up to balcony create an impossible object.

Sangria jugs of 75 l(itres) would be a bit big!

Children shown standing at the left-hand side of the pool where depth is 5 m(etres).

Depth shown as 10 cm where the diving boards are.

S2 **Four digits** (p 4)

This activity will tell you something about pupils' knowledge of number operations and symbols (for example, brackets) and their arithmetic skills. It also gives an opportunity for co-operative group work.

◊ Decide together on the four digits to be used. They do not have to be all different. 0 is not very helpful.

◊ You can allow free rein at first as far as the rules are concerned. Pupils will probably come up with some ground rules themselves, and then you can establish rules for everyone, for example:

- All four digits must be used.
- No digit can be repeated unless it occurs twice in the set.
- Digits can be used in any order.
- Any operations can be used. (Brackets may be needed and $\sqrt{}$ may be suggested by pupils.)
- Digits can be combined to make two- and three-digit numbers.
- Results must be whole numbers (for example, $43 \div (2 + 1) \neq 14$).

◊ You could start by asking for ways to make, for example, 10.

◊ Pupils could work in groups, each group making a collection. An element of competition could be introduced. Alternatively, groups could be given ranges of numbers (1–20, 21–40, ...).

◊ It is necessary to record the completed numbers and the methods used to arrive at them, for example, a list:

1	11	21
2	12	22
3	13	etc.
4	14	
5	15	
6 1 + 3 + 4 − 2	16	
7	17	
8 2 + 3 + 4 − 1	18	
9	19 12 + 3 + 4	
10 1 + 2 + 3 + 4	20	

◊ In one school the results were recorded on a large chart and put on the wall. This was added to over the year. (It works particularly well if there is some reward for completing gaps – merits, credits, etc.)

◊ Calculators may be used if necessary, although many pupils should do well without them.

◊ Pupils may realise that results often come in pairs, for example:
$$23 - 14 = 9, \quad 23 + 14 = 37$$
then with some swapping around:
$$32 - 14 = 18, \quad 32 + 14 = 46, \quad \text{etc.}$$

Follow-up Each pupil can choose their own set of four numbers. Some sets of numbers (for example, 6, 7, 8, 9) are more difficult than others and may lead to demotivation.

Two-piece tangrams (p 5)

These activities give an opportunity to use mathematical language for shapes, properties etc., and to encourage good presentation. Because the possibilities are limited in each case, it is fairly easy to find them all.

Essential	Optional
Squared paper, scissors, glue	Card Pre-drawn rectangles (see below)

T

1 **Rectangle** (p 5)

◊ The dimensions 7 cm and 4 cm are chosen because $\frac{7}{4}$ is approximately $\sqrt{3}$. An equilateral triangle (within the bounds of measurement error!) can be made with the two pieces.

◊ Instead of making several copies, pupils could cut one pair of triangles from card and draw round the shapes they make.

The pieces may be turned over to make the shapes.

◊ These are the different shapes that can be made:

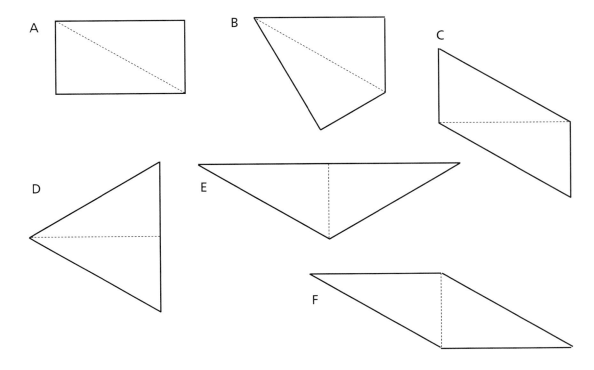

Follow-up You could write the names of the shapes on the board (kite, isosceles triangle etc.) and ask pupils to label their shapes.

Pupils could suggest ways of categorising. The results can be tabulated, for example:

	Sides	Name	Right angles	Lines of symmetry	Order of rotation symmetry
A	4	rectangle	4	2	2
B	4	kite	2	1	–

2 Square (p 5)

These shapes can be made:

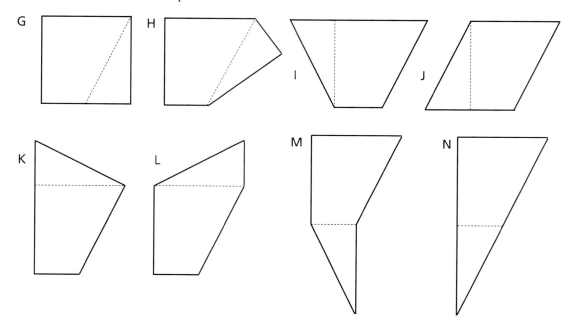

Extensions ◊ Pupils could start with their own shape and make a two-piece tangram.

◊ Add an extra cut to the rectangle or square to make a three-piece tangram. Suggestions (if needed):

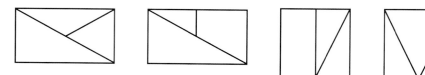

It is better if the cut gives equal lengths for joining.

Note The full seven-piece tangram is used in S9 'Tangrams'.

 **Lunch break** (p 6)

For many pupils, the entry to secondary school is the first time when they can choose from a lunch menu and pay themselves. This activity is intended to familiarise pupils with their own school's lunch menu and to give practice in mental arithmetic with money.

Essential	Optional
Copies of school's lunch menu, with prices	Coins

'Till staff were enthusiastic that students had a chance to work out their costs in advance.'

◊ Some schools found this worthwhile as part of their induction programme.

◊ You could begin by asking oral questions based on the menu, for example:
'How much does burger and chips cost?'
'If you buy a sandwich, how much change do you get from £1?'

8 Making the poster is likely to be more suitable for lower attainers. Posters could be displayed in Year 7 tutor rooms.

9 The last two puzzles are more demanding. They could be done by trial and improvement or by noticing that adding all three priced items together gives twice the unpriced item.

1–4 Answers depend on your school's prices.

5 £1.35

6 £1.65

7 The pupil's question and answer

8 The pupil's poster

9 (a) Bun, jam and cream: 64p
Bun and cream: 47p

(b) Samosa and relish: 86p
Onion bhaji and relish: 94p

(c) Pie, chips and onions: £1.90
Pie, chips, onions and beans: £2.52
Beans and chips: £1.29

(d) Egg, beans and tomato: £1.08

(e) Sausage, egg and chips: £1.97

S5 Finding your way (p 8)

This gives practice in using left and right and in reading a simple map.

A Left and right (p 8)

◊ Before discussing the picture and the questions, you could get a pupil to face the rest of the class, and ask which side of the room is on his or her left. ('Simon says' is also good for practice.)

◊ You could use a plan of your school. Give pupils a list of instructions from the classroom to somewhere else and ask them to guess the destination. They can then make up instructions for one another.

You could ask pupils to shut their eyes and imagine where they are going as you give them instructions.

◊ Before doing questions A12 to A15 you could do some practical work using the layout of the classroom. The aisles can be roads and the desks houses.

'Robots' is a good game for emphasising that instructions have to be precise. Initially you could be the robot and ask a pupil to give you instructions. Then pupils can take turns to work in pairs.

B Reading a map (p 11)

Similar work can be based on local maps.

A Left and right (p 8)

A1 Right

A2 Left

A3 Right

A4 Left

A5 Right

A6 Left

A7 Left

A8 Tasco car park, football ground

A9 Wood Hill, Elm Road and South Road

A10 The pupil's answer

A11 The pupil's answer

A12 (a) 2nd left (b) 1st right
(c) 3rd right (d) 5th left
(e) 4th right

A13 (a) Aspen Lane (b) Birch Grove
(c) 4th left

A14 Turn right into Beech Avenue, turn right when you get to Wood Street, take the 2nd right, Sam's is 4th house on the left.

A15 Turn right into Poplar Walk, turn left at Wood Street, take the 3rd right, Nikki's is 3rd house on the right.

B Reading a map (p 11)

B1 (a) Back Lane   (b) Downend Road
     (c) Left   (d) Right
     (e) Right   (f) Left

B2 Go along Windy Lane, take the 3rd turning on the left, the post office is on your left in Coronation Road.

B3 Turn left into Ashton Way, then turn right along Coronation Road, turn right, then 1st left, then 2nd left. Scott's is the house at the end of the road.

B4 Ashton Way and Mill Lane

B5 Robin Hall

B6 The pupil's journey

 **Gridlock** (p 13)

This game gives you an opportunity to find out pupils' addition (and subtraction) skills. Pupils can also develop and explain strategies to win.

Essential	**Optional**
Up to 15 dice for a class of 30	Sheet 47
	Sheets 48 and 49 (blank grids)

'I almost didn't use sheet 47 but it turned out to be most useful.'

◊ To help them understand the scoring system, pupils could complete the grids on sheet 47 and work out the scores.

Some schools have used this sheet for homework.

◊ Initially the class could play together, with you rolling the dice and calling the numbers. Then the game can be played in groups of two or more.

◊ More able pupils may think that the game is a trivial exercise requiring only simple addition skills. Emphasise early on that they should be thinking about good strategies to maximise their chance of winning.

To play 'Gridlock'

Each pupil draws a square grid (start with 3 by 3 grids) and marks off the top left-hand section as shown.

The caller rolls a dice and calls out the number. Each pupil writes the number in any empty square in the section shown shaded on the right.

Repeat until each square in that section is filled.

Each number must be written in the grid before the next is called and a number can't be changed once it is written.

Each pupil adds up their numbers in the rows, columns and diagonal and writes the totals in the empty squares as shown on the right.

Each pupil adds up their points.
- Score 2 points for a total that appears twice.
- Score 3 points for a total that appears three times.
- Score 4 points for a total that appears four times ... and so on.

The grid above scores 4 points (6 and 7 both appear twice as a total).

After a number of rounds (decided by you), pupils add up their points and the one with most points is the winner.

◊ After playing on 3 by 3 grids, play the game on larger ones.

◊ Once pupils have played the game a few times, ask them to describe any strategies they use in placing the numbers on their grids. For example, if a number is rolled twice it is better to place the numbers diagonally,

for example [5 in top-left, 5 in bottom-right] or [5 in top-right, 5 in bottom-left] rather than [5 5 in top row]

◊ Now you can alter the rules as follows. First, the numbers called out are written at the side of the grid. When all numbers have been called, they are then placed in the grid. Pupils can think about how to get the maximum possible score with a particular set of numbers.

◊ One variation is for the winner to be the person with the fewest points. Pupils can discuss how their winning strategies change in this case.

Another variation is to use two dice to generate larger numbers.

Follow-up In questions 1 to 11, the later questions are more difficult.

Remind pupils that they can only use numbers on an ordinary dice (1 to 6) to solve these problems.

1 In part (b), emphasise that their problems should be able to be solved without any guesswork or mind reading! Encourage more able pupils to make up problems that give the minimum necessary information.

This could be set as a homework task.

◊ You could ask pupils how changing the system of scoring points would affect the game, for example:

• score 2 points for a total that appears twice

• score 4 points for a total that appears three times

• score 6 points for a total that appears four times ... and so on

◊ A different version of the game is for pupils to cross out any totals that repeat and to add the remaining totals to give their score for that round. The winner could be the person with the most or fewest points.

1 (a)

2	**3**	5
3	**4**	**7**
5	7	**6**

4	**1**	5
3	6	9
7	7	**10**

Points scored: **4** Points scored: **2**

1	6	**5**	12
5	**3**	1	9
6	**4**	6	16
12	13	**12**	10

Points scored: **3**

(b) The pupil's problems

2 Examples of grids that score 2 points are:

5	3	8
6	2	8
11	5	7

3	6	9
5	2	7
8	8	5

3 Examples are:

(a)

6	6	12
4	3	7
10	9	9

(b)

4	6	10
6	3	9
10	9	7

4 (a) 2 points (b) 0 points

5 (a)

4	**2**	6
2	**1**	3
6	3	5

(b)

3	**1**	4
2	**5**	7
5	6	8

(c)

5	**2**	**3**	10
4	6	**4**	14
1	**1**	**1**	3
10	9	8	12

6 Examples are: 1, 2, 3 and 5; 2, 3, 4 and 6.

7 The pupil's explanation

8 The pupil's explanation

9 (a)

6	**4**	10
4	**3**	**7**
10	7	**9**

(b)

1	**5**	**6**
4	**6**	10
5	11	7

10 Examples are:

(a)

1	2	6	9
2	3	5	10
6	5	4	15
9	10	15	8

(b)

1	2	5	8
2	3	6	11
5	6	4	15
8	11	15	8

(c)

5	1	2	8
5	2	3	10
4	6	6	16
14	9	11	13

11

1	**5**	**5**	11
1	**3**	**4**	8
6	**5**	**6**	17
8	13	15	10

Sheet 47

1

6	5	**11**
1	1	**2**
7	**6**	**7**

Points scored: **2**

2

3	4	**7**
6	5	**11**
9	**9**	**8**

Points scored: **2**

3

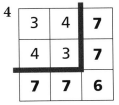

3	5	**8**
2	4	**6**
5	**9**	**7**

Points scored: **0**

4

3	4	**7**
4	3	**7**
7	**7**	**6**

Points scored: **4**

5

6	1	2	**9**
3	4	6	**13**
1	5	1	**7**
10	**10**	**9**	**11**

Points scored: **4**

6

5	3	5	**13**
3	1	4	**8**
2	6	1	**9**
10	**10**	**10**	**7**

Points scored: **3**

 **Year planner** (p 15)

This activity teaches, or reinforces, basic information about the calendar, in particular the number of days in each month. In some trial schools, it proved to be more difficult for pupils than initially anticipated.

Essential
Sheet 50

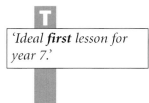

'Ideal **first** lesson for year 7.'

'It was not obvious to the students where to start the months.'

◊ Discuss the rhyme on page 15 and make sure that pupils are clear about the number of days in each month. Pupils may be able to suggest other ways of remembering them.

You may wish to discuss the idea of a leap year and the reasons for an extra day in February every four years. This could lead to a discussion on how to test if a year is divisable by 4.

◊ Find out on what day of the week 1 September falls.
Pupils write '1' in the first such day on the planner for September. They then number the following days to 30.

If 30 September falls on, say a Wednesday, then 1 October will fall on a Thursday. Pupils enter a '1' in the first available such day in the planner for October. Some pupils may find this quite challenging. They continue until every month is filled in.

◊ They can use the planner to record (for example, by shading) school holidays and any particular dates of importance or interest. A key could be used to indicate different activities. For example, red could be used for holidays, blue for birthdays, green for religious festivals etc. Individual planners could then be stapled into school day books, diaries etc. and/or a large copy made for wall display.

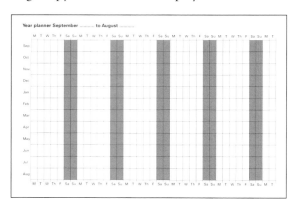

Patterns from a hexagon (p 16)

This work is to help pupils develop skills with compasses and rulers that are needed later to construct triangles and angles. Pupils also analyse patterns and make decisions about how to construct them.

At the start of the year many pupils have coloured pencils and geometry sets, so capitalise on this.

Essential

Sharp pencils
Pairs of compasses
Rulers
Coloured pencils
Board compasses

'I discovered that only about half the class had used compasses before.'

◊ Many pupils find it difficult to draw a circle with a pair of compasses. They may need to draw circles and simple patterns before they feel confident enough to try the more difficult patterns.

Many pupils will find it helpful to see a demonstration of how to draw a regular hexagon. They must be able to draw a regular hexagon in order to draw the patterns on page 17.

Common problems include

• not realising that the point of the compasses is moved to the point where the last arc crosses the circumference (and not at the end of the arc) for subsequent arcs to be drawn

• not realising that, to draw the hexagon, you join points where the arcs cross the circumference (and not the ends of the arcs)

Although pupils may have drawn them before, it may be helpful to demonstrate on the board or OHP how to draw the seven-circle or petal designs shown below.

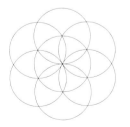

◊ This work provides good material for wall displays. In one school, the hexagon designs were used to make mobiles.

◊ You may want pupils to leave construction lines so you can check their methods.

◊ The construction of the patterns on page 18 offers more of a challenge, and the construction gets more involved further down the page.

The last two designs can be drawn as follows:

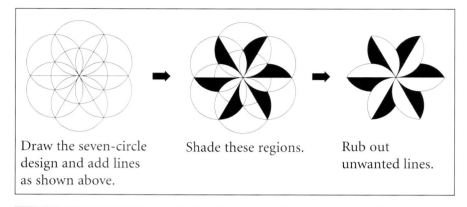

Draw the seven-circle design and add lines as shown above.

Shade these regions.

Rub out unwanted lines.

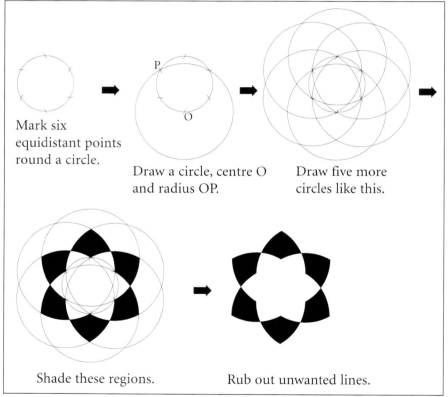

Mark six equidistant points round a circle.

Draw a circle, centre O and radius OP.

Draw five more circles like this.

Shade these regions.

Rub out unwanted lines.

Follow-up If they know about symmetry, pupils could try to draw a pattern with 0 lines of symmetry, 1 line of symmetry etc.

Tangrams (p 19)

This develops spatial awareness and includes comparison of areas using a non-standard unit.

Essential
Sheet 51 (print on coloured card, one copy per six pupils)
Sheets 52, 53, 54, 55

◊ If pupils have difficulty making the shapes, you can show them where one of the largest triangles goes.

◊ The tangram is often described as an ancient Chinese pastime. However, although Chinese in origin, it dates from about 1800. There are several books of tangram puzzles available. Some tangram sets are made with magnetic pieces.

The history and mathematics of tangrams are explored in a number of books on recreational mathematics.

1 Sheet 52

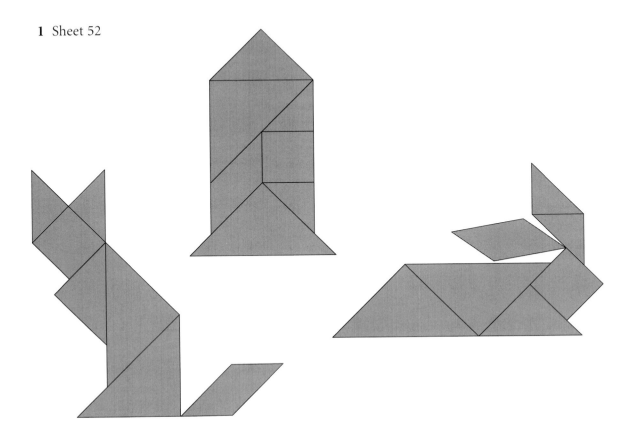

2 The pupil's own tangram

3 (a) 4   (b) 2   (c) 2   (d) 2   (e) 16

4 Sheet 53

(a) 10   (b) 8   (c) 9   (d) 8

Sheet 54

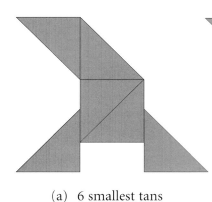

(a) 6 smallest tans

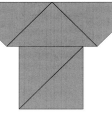

(b) 5 smallest tans

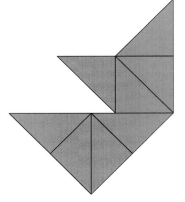

(c) 8 smallest tans

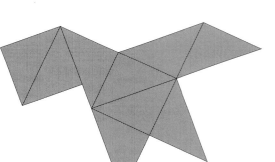

(d) 9 smallest tans

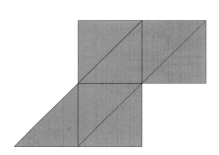

(e) 7 smallest tans

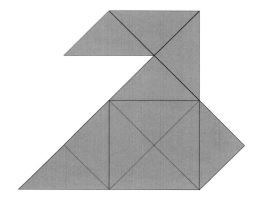

(f) 11 smallest tans

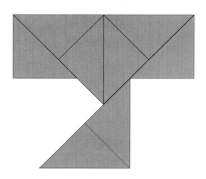

(g) 8 smallest tans

5 At a glance it looks as though the second shape on sheet 55 has a smaller area, because of the 'missing piece'. However, both shapes are made from a complete set of seven tans and therefore have the same area.

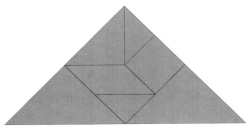

Area is 16 smallest tans

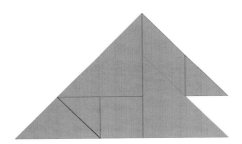

Area is 16 smallest tans

*6 A: 3 smallest tans
   B: 2 smallest tans (the difference between a large triangle and a medium triangle)
   C: 4 smallest tans (remove one smallest tan to make the area of the hole 5 smallest tans)
   D: 2 smallest tans (the difference between a large and a medium triangle)

 **Half a square** (p 21)

This activity will help you find out what pupils know about area, symmetry and congruence of shapes.

Essential	Optional
Squared paper	Triangular dotty paper

T

◊ If you cut squared paper into small pieces and pupils draw one 4 by 4 square on each piece, it is easier for them to sort and display their diagrams.

◊ You could start by drawing some squares on a grid on the board, each with a line across and part shaded. Ask the class to vote on whether each line divides the square into halves or not. To check, you could get a pupil to shade a section equivalent to the already shaded section.

Ask pupils to split the square so that each half has the same area. Discuss the strategies they use to ensure that the two areas are equal.

Some pupils may limit themselves to congruent shapes and check by tracing.

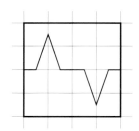

'This activity has masses of potential for extension. We found that after pupils have drawn their own diagrams (perhaps for homework), a follow-up lesson is most necessary to make the exercise worthwhile.'

Others might 'take a bit out and put it back':

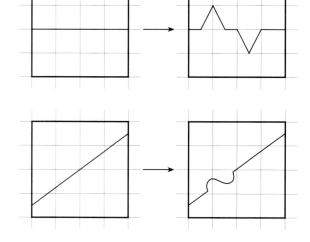

◊ You could ask different groups of pupils to tackle the task in different ways. What if they
  • draw one straight line
  • start and finish on opposite sides of the square
  • start and finish on adjacent sides of the square
  • start and finish on the same side of the square

◊ You could introduce the word **congruent**. Reflection and rotation symmetry could also be introduced for some pupils.

◊ Ask the pupils to collect together designs that have a common property. This is a good opportunity to encourage pupils to use their own ideas first. Discuss the properties that they use.

◊ Ask pupils to collect together designs which have
  • line symmetry
  • two congruent halves
  • rotational symmetry
  Discuss why any designs with rotational symmetry must have congruent halves.

◊ Pupils could then look at a set of designs with a common property and try to
  • decide what the common property is
  • draw another design for that set

Extensions Pupils could consider ways of splitting other shapes in half, for example:

equilateral triangles

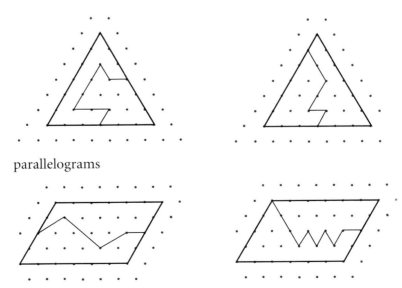

parallelograms

Pentominoes (p 22)

Essential	Optional
Sheets 56 and 57	Square tiles (about 60 for each pair of pupils) Tracing paper

T

Ⓐ Finding pentominoes (p 22)

> Optional: Sheet 56 (at least one for each pair of pupils), square tiles (about 60 for each pair of pupils), tracing paper

◊ Pupils could start with shapes made from three squares and four squares (trominoes and tetrominoes).

◊ After showing pupils what a pentomino is ask them to find all possible different pentominoes. There is likely to be some discussion on the possible meanings of 'different' and 'same' here. 'Different' is usually taken to mean non-congruent. Tracing paper helps pupils identify pentominoes that are the same. Tiles are useful to make the pentominoes.

The twelve pentominoes are shown below with possible labels:

> *'They **all** worked hard to find the twelve pentominoes – a Mars bar was at stake! The concept of same shape/different position came slowly to some but eventually they all understood.'*

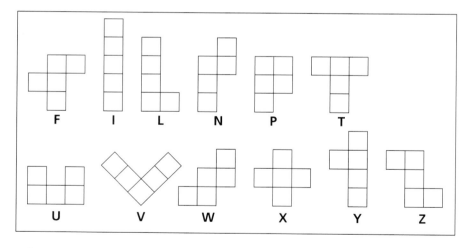

◊ Ask pupils how they can be sure they have found them all. Encourage them to be systematic. For example, they could find all those with five squares in a row, then four in a row, then three in a row and finally two in a row.

◊ When pupils think they have found all twelve pentominoes they could compare their results with the set on sheet 56.

◊ An additional activity is to fit all twelve pentominoes together to make a rectangle with no gaps.

There are 3719 solutions but don't expect any to be found quickly! The number of solutions for each possible size of rectangle is

6 by 10   2339 solutions
5 by 12   1010 solutions
4 by 15   368 solutions
3 by 20   2 solutions (very difficult)

◊ One school used the pieces as the basis of some work on area, perimeter, reflection and rotation symmetry. The perimeter of each shape was found to discover which shapes had the smallest/largest perimeter. The shapes were then classified as having reflection or rotation symmetry. The pupils found it useful to be able to physically turn and fold the shapes.

B The 8 by 8 game (p 23)

This activity gives pupils an opportunity to consider strategies for winning a game played with a set of pentomino pieces.

> Essential: Sheets 56 and 57 (one of each for each group)

◊ The game is best played in groups of two or three.

◊ Encourage pupils to consider strategies for winning. For example, with two players the second player should try to move so that there is room only for an even number of pieces.

◊ One extension activity is to try to place all twelve pentominoes on the 8 by 8 board with four squares left over. This is difficult but solutions exist for all positions of the four squares.

◊ Another extension is to try to place pentominoes on the board in such a way that none of the rest can be added. The minimum number is five.

C Pentomino fields (p 24)

These pentomino problems give an opportunity to compare areas by counting squares.

> Essential: Sheet 56
> Optional: Sheet 57 (at least two for each pair)

◊ Pupils can use two copies of sheet 57 to make a grid of squares that is large enough for each solution. The areas can be found more easily.

C **Pentomino fields** (p 24)

Solutions showing the maximum possible field areas are:

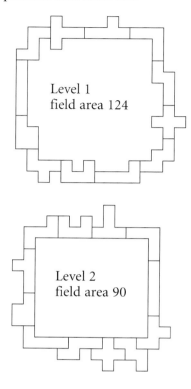

Level 1
field area 124

Level 2
field area 90

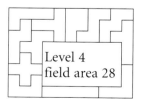

Level 3
field area 61

Level 4
field area 28

Martin's solution to the level 2 problem:

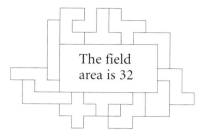

The field area is 32

Mary's solution to the level 3 problem:

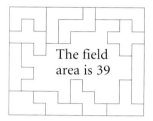

The field area is 39

Ellen's solution to the level 4 problem is the maximum one shown on the left.

Shapes on a dotty square (p 25)

Pupils create shapes and use mathematical language to describe them.
They can also decide on their own lines of investigation.

Optional

Square and triangular dotty paper
Sheet 58
3 by 3 pinboards, rubber bands
Tracing paper
OHP transparency of square dotty paper

'For some pupils, drawing out the grid was the hardest part.'

◊ Square dotty paper can be used or pupils can draw grids of dots on square paper. Sheet 58 is for those who find it difficult to 'rule off' the grids.

◊ Establish rules for drawing shapes on the pinboard/grid.

- Only the 9 pins/dots can be used.

- All corners must be at a pin/dot.

- The types of shape shown in the pupil's book are disallowed (ones with 'crossovers' or 'sticking out' lines).

◊ Ask pupils to draw a few different shapes following the rules above.

There is likely to be some discussion on the possible meanings of 'different' and 'same' here. 'Different' is usually taken to mean non-congruent. Tracing paper helps pupils identify shapes that are the same.

Look at some of their shapes together.

'Computers helped resolve arguments as to "sameness" of shapes by rotating, reflecting and superimposing.'

- What properties have they got?
  (Number of sides, angles, symmetry, parallel sides, area, …)

- Do pupils know names for any of the shapes?
  (Triangle, quadrilateral, rectangle, parallelogram, hexagon, …)

◊ There are various ways to structure this activity. Some trial schools generated a collection of questions from which pupils chose. For example:

What shapes have the most sides?
How many different squares?
How many different triangles?
How many shapes have reflection symmetry?
How many shapes have rotation symmetry?
What different areas can you make?
How many ways can you put the same triangle on the grid?
What shapes can be made with 1, 2, 3, … right angles?

Pupils can choose a question (or pose one of their own) and write up their solution. These could then be displayed.

For example, the 23 different polygons with reflection symmetry are:

◊ An alternative structure is to begin by asking how many different triangles can be found. The 8 different triangles are:

Now discuss the properties of the triangles. For example:

Which has the greatest area?
Which has a right angle?
Which has reflection symmetry?
Which are isosceles?

Pupils can now consider the different quadrilaterals that can be found and their properties. The 16 different quadrilaterals are:

Pentagons, hexagons and heptagons can be considered but the number of different shapes may be rather daunting!

The numbers of different polygons of each type are:

Number of sides	Number of different polygons
3	8
4	16
5	23
6	22
7	5

This gives 74 different polygons.

Extension ◊ One extension is to consider polygons on a different grid. Suggestions appear in the pupil's book.

The hexagonal 7-pin grid yields 19 different polygons which is a manageable number for most pupils to find. The polygons are:

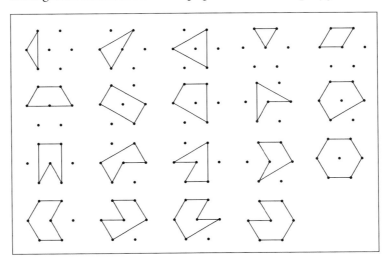

Reflection symmetry

Pupils use folding/cutting and mirrors in a variety of practical activities. They have especially enjoyed the symmetry tiles game (section E) and it has proved very effective in consolidating the key ideas.

Rotation symmetry is not dealt with in this unit. However some pupils are likely to be aware of the idea and it is there as a 'distractor' in some questions. You may wish to discuss it where appropriate.

There is more material here than one pupil could be expected to do in the time likely to be available. Once you have seen how much pupils already know about symmetry, you can choose appropriate work.

Essential	**Optional**
Mirrors	Tracing paper
Scissors	
Paper for folding and cutting	
Square dotty paper or cm squared paper	
Sheets 59 to 66	
Practice sheets P49 to P57	

Ⓐ Folding and cutting (p 26)

Pupils predict a mirror image and check by folding and cutting (including reflecting in sloping lines).

Scissors, sheets 59 (▼■●) (could be omitted by high attainers), 60 (▼■○), 61 (▼■●), 62 (▽■●)

Optional: Mirrors/tracing paper (for pupils using alternative methods)

'As an introduction, pupils drew the 2nd half of a mask. They found it fun and the masks made a good display. They knew what to do in a non-mathematical context. This built confidence.'

◊ For the introductory folding and cutting activity, higher attainers could use sheet 61 and omit sheet 59.

◊ Pupils could use alternative methods (a mirror and/or tracing paper) but folding and cutting gives an immediate check that is very convincing. You may wish to allow a variety of methods in the classroom depending on pupils' prior attainment.

◊ Sheet 61 should reveal the common errors when reflecting in sloping lines. Counting dots outwards from the mirror line can be a useful approach.

Ⓑ Using a mirror (p 27)

Pupils use a mirror to draw and check a mirror image (including reflecting in sloping lines).

Mirrors, sheet 63

◊ It may be worth leading the class through the stages in the photographs, stressing the meaning of the word 'image'.

Ⓒ Looking for reflection symmetry (p 28)

Pupils identify lines of symmetry and use a mirror to check.

Mirrors

◊ It may be worth leading the class through the stages in the photographs.

C2 Some pupils may say that designs (d), (g), (j) and (l) have reflection symmetry because they recognise the rotation symmetry. It is a good opportunity to discuss this type of symmetry.

Some may say that (p) has reflection symmetry because they can see that the 'cut out' shape on the left is the same as the protruding shape on the right. This is an opportunity to discuss translation.

Ⅾ **Times and dates** (p 30)

Pupils find lines of symmetry in groups of digits (times and dates) where two lines of symmetry are possible.

> Mirrors

◊ Many pupils find it harder to identify a horizontal line of symmetry than a vertical one. It may be worth leading a brief discussion after D1 has been completed.

◊ Some pupils may not be familiar with the method of writing dates used in D2 onwards. It may be beneficial to discuss this with pupils before they begin this work.

'Surprisingly, pupils were not proficient at putting dates into figures.'

D4 Pupils may be uncomfortable about the answer 'none' for (c) as they may recognise the rotation symmetry. It is a good opportunity to discuss this type of symmetry.

◊ Emphasise that dates found in D6 and D7 have to be possible.
For example, 83:38:83 is not a possible date!

ⅇ **Symmetry tiles** (p 31)

Pupils consolidate work on reflection symmetry by using tiles to make symmetrical patterns. The tiles are also used to play a game.

> Scissors, mirrors, sheet 64 (copied on card if possible),
> sheet 65 (one for each group of players)

◊ In problems E1 to E6 there is no need for pupils to draw diagrams to show their results but some may wish to do so.

Symmetry tiles game

◊ The game should be self-correcting: hopefully, players will protest at an invalid move. Some care is needed in organising the groups: each group should have one pupil who is confident enough about symmetry to recognise invalid moves. Although the game can be played with four players, having only two or three makes it faster and more enjoyable.

◊ You may have to clarify one or two things: pupils can put their cards down on either side of the dotted line; they don't have to complete the symmetry at every move (though they could play that way).

◊ Watch for pupils who are still getting the symmetry wrong. Provide them with a mirror so they can see their mistakes.

F **Folding again** (p 34)

Pupils predict and make shapes with two lines of symmetry by folding and cutting. They also find lines of symmetry for the letters of the alphabet.

> Scissors, loose sheets of paper (newspaper might be useful), sheet 66
>
> You could prepare some shapes on 'twice-folded paper' such as the ones at the top of page 34 to use in your introduction.

One way to structure the introduction is as follows.

◊ Fold a sheet of A4 in half and half again like this.
Make sure pupils can see where the folds are.
(Draw along each fold with felt-tip pen if there is any doubt.)

Draw on the paper, for example like this.
Ask what the shape will be like when you cut it out and unfold it.

Unfold the sheet for pupils to see.

◊ Now show the three examples at the top of page 34.
Ask pupils which of them will make a letter of the alphabet when cut and opened out. Test their prediction by cutting and opening out.
Ask pupils to describe what the other shapes will look like when they are cut and opened out. Test their predictions by cutting and opening out.

F3 This is a challenging question that is likely to provoke discussion. Encourage pupils to think about the lines of symmetry and how they relate to the shape to be drawn.

Newspaper could be useful here.

F4 The 'O' on sheet 66 is a circle and so has an infinite number of lines of symmetry.

G **Rangoli patterns** (p 36)

Pupils make traditional Sikh and Hindu patterns.
The patterns have four lines of symmetry.

> Squared or dotty paper
> Optional: Mirrors/tracing paper

◊ Remind pupils that any colouring should be symmetrical too.

H **Shading squares** (p 38)

Pupils make designs with one or more lines of symmetry by shading squares. They can try to devise strategies to ensure all the different ways of shading squares are included.

> Optional: Mirrors (to check results)

◊ Lower attaining pupils may feel more confident if they are told there are four ways for each problem in H1 and H2.

◊ Encourage higher attainers to explain how they know they have found all the different ways for each problem.

◊ These problems can be extended in a variety of ways. In one school, pupils suggested their own extensions and looked at shading different numbers of squares on these diagrams.

I **Designs** (p 39)

Pupils decide if a line is or is not a line of symmetry.

> Optional: Mirrors (to check results)

Ⓐ Folding and cutting (p 26)

A1 to **A5** The pupil's drawings and checks

Ⓑ Using a mirror (p 27)

B1 to **B2** The pupil's symmetrical drawings

Ⓒ Looking for reflection symmetry (p 28)

C1 (b) Yes (c) Yes (d) No (e) Yes
(f) No (g) Yes (h) Yes

C2 (a) Yes (b) No (c) Yes
(d) No (e) Yes (f) No
(g) No (h) Yes (i) Yes
(j) No (k) Yes (l) No
(m) Yes (n) No (o) Yes
(p) Yes

Ⓓ Times and dates (p 30)

D1 (a) 01:18, 03:38, 11:18, 01:00, 13:13, 01:10 and 10:01 are symmetrical.
(b) 01:10 and 10:01 have two lines of symmetry.
(c) The pupil's three times with one line of symmetry
(d) One time from 11:11 and 00:00

D2 Two lines of symmetry

D3 Yes, it is symmetrical.

D4 (a) None (b) Two (c) None
(d) One

D5 (a) 04:02:33 none
(b) 31:10:81 one
(c) 08:11:80 two
(d) 08:01:80 one

D6 The pupil's two dates with one line of symmetry

D7 The pupil's two dates with two lines of symmetry

Ⓔ Symmetry tiles (p 31)

E1 and **E2** The pupil's patterns

E3

E4 In each case, there is another solution with the tiles on the other side of the dotted line.

(a) 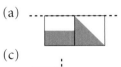   (b)

(c)

E5
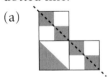

E6 In each case, there is another solution with the tiles on the other side of the dotted line.

(a)   (b)

E7 In each case, there is another solution with the tiles on the other side of the dotted line.

(a)   (b)

(c)   (d)

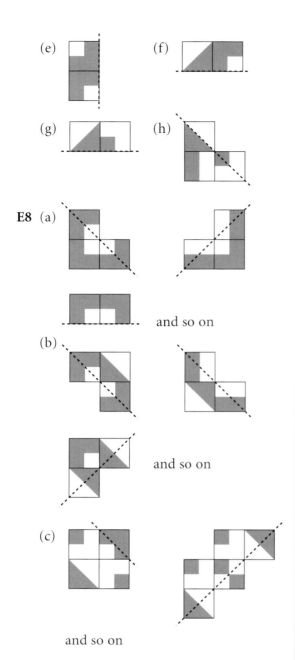

(e)

(f)

(g)

(h)

E8 (a)

and so on

(b)

and so on

and so on

(c)

and so on

F Folding again (p 34)

F1 (a) Shape Q (b) Shape Y

(c) The pupil's checks

F2 Shape B makes I and shape C makes X.

F3 The pupil's attempts to produce the shapes

F4 (a) The first letter with no line of symmetry is F.

(b) A few lines of symmetry are drawn on 'O' as examples: 'O' has an infinite number of lines of symmetry.

(d)

Number of lines of symmetry	Letters
0	F G J L N P R S Z
1	A B C D E K M Q T U V W Y
2	H I
3	
4	X
more than 4	O

G Rangoli patterns (p 36)

G1 (a) Two lines of symmetry

(b) Four lines of symmetry

G2 The pupil's drawings

H Shading squares (p 38)

H1 Four different ways

H2

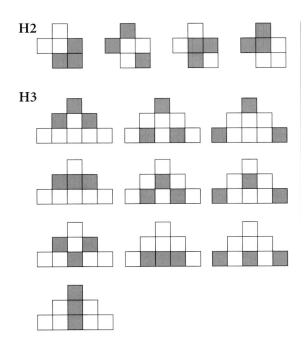

H3

H4 Three different ways

H5 Four different ways

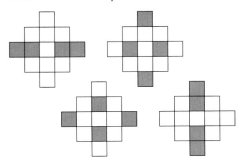

Designs (p39)

I1 (a) Lines 1 and 3

(b) Lines 1, 2, 3 and 4 (all of them)

(c) Lines 1, 2, 3 and 4 (all of them)

(d) Neither line is a line of symmetry.

(e) Lines 1 and 3

(f) Lines 1, 3 and 5

What progress have you made? (p 40)

1 (a) 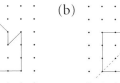 (b)

2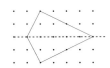

3

4 Lines 2, 4 and 6 are lines of symmetry.

5

Practice sheets

Sheet P49 (section A) ▼□○

1 The pupil's drawings and checks

Sheet P50 (section A) ▽■●

1 The pupil's drawings and checks

Sheet P51 (section B) ▼□○

1 The pupil's symmetrical drawings

Sheet P52 (section B) ▽■●

1 The pupil's symmetrical drawings

Sheet P53 (section C)

1 (a) Yes (b) Yes (c) Yes (d) No
 (e) No (f) No (g) Yes (h) No

Sheet P54 (section C)

1 (a) Yes (b) Yes (c) Yes (d) No
 (e) Yes (f) No (g) No (h) Yes
 (i) Yes (j) No (k) Yes (l) Yes

Sheet P55 (section D)

1 1881, 1331, 1301, 1811 and 1380 are symmetrical.

2 The pupil's four symmetrical years

3 (a) 1881 has two lines of symmetry.

 (b) The pupil's two years with two lines of symmetry: for example, 1001, 8008, 8118, 1111, 8888

4 (a) One line   (b) No lines
 (c) Two lines   (d) One line

5 (a) 130 (b) 380 (c) 99 (d) 121
 The answers to (a) and (b) are symmetrical.

6 The pupil's symmetrical sums

Sheet P56 (section D)

1 1881, 1331, 1301, 1811, and 1380 are symmetrical.

2 1800, 1801, 1803, 1808, 1810, 1811, 1813, 1818, 1830, 1831, 1833, 1838, 1880, 1881, 1883, 1888

3 (a) 1881 has two lines of symmetry.

 (b) 1, 8, 11, 88, 101, 111, 181, 808, 818, 888, 1001, 1111, 1881

4 (a) One line   (b) No lines
 (c) One line   (d) One line

5 The pupil's multiplications with answers that are symmetrical
 (The answer to (a) in question 4 is symmetrical.)

Sheet P57 (sections H and I)

1 (a) Lines 1, 2, 3, 4 and 5 (all of them)

 (b) Neither is a line of symmetry.

 (c) Lines 1 and 5

2 (a) Four ways

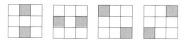

 (b) Four ways

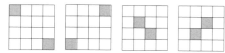

 (c) Eight ways

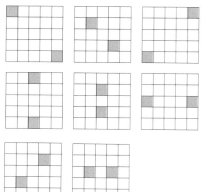

Test it!

Pupils collect measurements to test general statements.
Higher attainers go on to find an approximate relationship between two
sets of measurements.

T

p 41	**A** I don't believe it!	▼■●	Planning a task Collecting data to test a statement Measuring
p 42	**B** Organising your results	▼■●	Organising measurements
p 44	**C** Now it's your turn!	▼■●	Testing a chosen statement
p 44	**D** Number detective	▽□●	Using division to find an approximate connection between two sets of measurements

---

Essential

Metre sticks, tape measures (enough for at least one item per group)

A I don't believe it! (p 41)   ▼■●

Metre sticks, tape measures (at least one item per group)

◊ Organise pupils into small working groups. (Groups of four work well.)

After you have introduced the statement 'Everyone is six and a half feet tall', the groups can discuss the first set of questions.

It should become clear that

- six and a half feet tall means six and a half 'foot lengths' tall
- the statement can be tested by measuring or simply stepping off each pupil's foot length against their height

You could have a general discussion at this point comparing the groups' plans for testing the statement. Or the groups could move straight into carrying out their plans. Some pupils may need a lot of help with measuring.

Recording of data may be haphazard. This is taken up in section B, which some teachers have preferred to do before A.

Methods used by pupils include:

- making an outline of themselves on a roll of old wallpaper, then cutting out their footprints to test the statement
- Blu-tacking rulers to walls to make it easier to measure heights
- measuring out heights on the tape and then 'stepping off'

Six and a half feet tall

In work of this kind it is important to make a plan and to adapt it as necessary. In many cases pupils do not see the need for a plan, preferring instead to 'jump right in'. You may find examples to emphasise this point in the pupils' own work.

Encourage each group to compare findings with others.

◊ After this discussion it is worth raising the issue of whether the statement 'Everyone is six and a half feet tall' is true for a wider population. Pupils could investigate the statement by measuring younger or older people at home for homework.

'Pupils enjoyed this topic. However, there were several teething problems such as which groups they were in and who was responsible for what. When things settled down they produced some excellent work which was good for display. The groups also gave a presentation of their work.'

B Organising your results (p 42)

◊ Some teachers have preferred to do this section before section A.

B1 Pupils should consider the problems of
- mixed units
- writing results in different orders

B2 You may need to help with 'approximately'.

C Now it's your turn! (p 44)   ▼■●

Pupils choose their own general statement to test.

◊ Each group must decide how to measure, for example, the 'length' of a person's head.

◊ A formal write-up is not necessarily expected at this stage. You could ask for a poster from each group or each pupil. (Later in the course there is a more specific focus on writing up results.)

Ⓓ Number detective (p 44)   ▽ □ ●

Pupils try to find an approximate relationship between two sets of measurements.

◊   This is suitable for pupils working in pairs.

D2   Some pupils may discover that 1.9 is a slightly better multiplier than 2.

D3   If pupils suggest other relationships, by all means let them investigate.

Ⓑ Organising your results (p 42)

B1   This method can be improved by presenting results in the same order and everyone using the same units.

B2   Ben is correct.

B3   (a)   It may mean the height from the chin to the top of the head (when the mouth is closed).

   (b)   Tim's arm span might be 170 cm.
   Gina's height might be 1.44 m.
   Sue's height might be 1.59 m.
   Sue's foot length might be 26 cm.
   Ryan's height might be 1.5 m.
   Lara's hand span might be 17 cm.

   (c)   Tim, Sue, Ryan (if he is 1.5 m tall) and Majid can go on the rides.

   (d)   Neena's arm span might be about the same length as Ajaz's, 153 cm.

   (e)   Depends on the pupil's own measurements

   (f)   About 6.4 to 7.3 times
   (The height and head length have to be expressed in the same units before dividing.)

Ⓓ Number detective (p 44)

D1   (a)   4

   (b)

Height (cm)	Elbow distance (cm)	Height ÷ arm
160	40	4
136.5	35	3.9
175.5	45	3.9
161	41	3.92…
133	35	3.8
152	42	3.61…

   (c)   The height is roughly four times the distance from elbow to finger tip.

D2   The approximate rule is that the height is twice the distance round foot.

D3   The pupil's own work

What progress have you made? (p 45)

1   The height is about 9 times the hand length.

③ Coordinates

<table>
<tr><td>

Essential

Sheet 67
Sheets 68, 69 (alternatives)

Practice sheets P58 to P60

</td><td>

Optional

OHP transparency of sheet 67

</td></tr>
</table>

Ⓐ **Recording positions** (p 46)   ▼■●

> Sheet 67 (an OHP is also useful)

◊ The two class games described below can be played at any time.

Coordinate bingo

Each pupil draws a grid labelled 0 to 5 on each axis. They mark seven grid points with little circles. This is their bingo card. You have a grid as well and call out points at random.

Four in a line

Draw a grid on the board labelled 0 to 6 on each axis. Choose two players who take turns to say the coordinates of a point. Label the points with the players' initials. The first to get four of their points in a line is the winner.

For a more demanding, but more interesting, game make it five in a line on a 0 to 9 grid.

Ⓑ **Digging deeper** (p 49)

> Sheets 68 (whole-number coordinates), 69 (includes $\frac{1}{2}$ and 0.5)

◊ Question B1 is for practice in whole-number coordinates.
Pupils who do not need this can start at B2.

Ⓒ **Negative coordinates** (p 51)

◊ The raised negative sign helps to distinguish between the number ⁻3 (negative 3 or minus 3) and the operation − 3 (subtract 3).

◊ You could play 'Coordinate bingo' again, with a grid labelled from ⁻3 to 2.

Ⓐ Recording positions (p 46)

A1 (a) Boy's boot (b) Silver brooch
  (c) Wooden comb

A2 (a) (3, 7) (b) (7, 10) (c) (6, 3)

A3 A wall **A4** (0, 0)

A5 A (3, 3), B (8, 8)

A6 Sheet 67

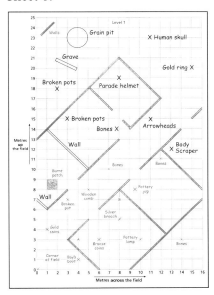

Ⓑ Digging deeper (p 49)

B1 Sheet 68

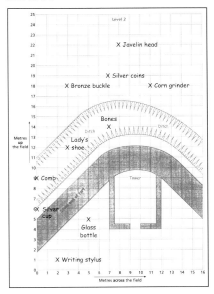

B2 Sheet 69

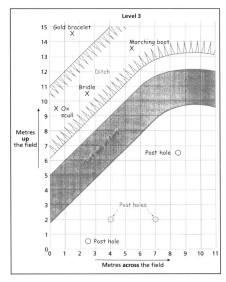

B3 Sheet 69

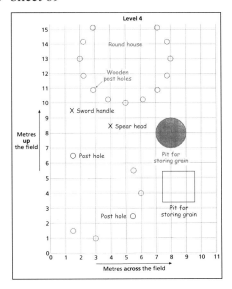

ℂ Negative coordinates (p 51)

C1 (a) $(^-2, 2)$ (b) $(4, ^-2)$ (c) $(^-2, ^-3)$
    (d) $(2, ^-4)$ (e) $(^-4, 0)$ (f) $(^-6, 2)$
    (g) $(0, ^-3)$

C2

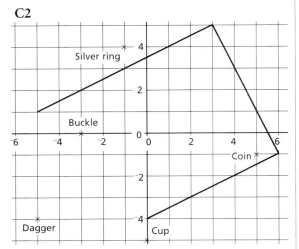

C3 (a) $(^-1, 4), (^-1, ^-2), (5, ^-2)$

     (b) $(^-1, 1), (2, ^-2)$

C4 $(4, 0), (2, ^-1), (0, ^-1), (^-2, 0), (^-2, 2)$

C5 (a) $(7, 2), \ (1, 8), \ (^-6, 1), \ (0, ^-5)$

     (b) $(3.5, 1.5), \ (1, 5), \ (^-2.5, 1.5), \ (0, ^-2)$

*C6 $(58, 39), (61, 39), (61, 41), (58, 41)$

What progress have you made? (p 53)

1 A $(0, 3)$, B $(4, 1)$, C $(6, 0)$

2

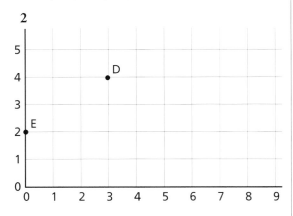

3 The points $(^-7, 2)$, $(^-3, ^-6)$ and $(1, ^-2)$
plotted on a grid

Practice sheets

Sheet P58 (section A)   ▼■●

1 (a) SMILE   (b) WITH
    (c) YOUR   (d) FRIENDS

2 (a) $(3, 1)$ $(0, 4)$ $(4, 1)$ $(0, 0)$ $(2, 1)$
      $(3, 4)$ $(0, 4)$ $(4, 0)$
    (b) $(3, 2)$ $(3, 3)$ $(1, 3)$ $(2, 3)$ $(4, 1)$

3

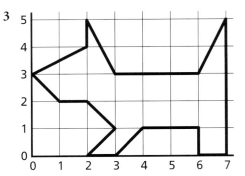

Sheet P59 (section B)   ▽■●

1 (a) $(0, 2), (3, 4), (6, 2), (3, 0)$
    (b) $(3, 2)$

2 (a) $(4, 5), (4, 8), (7, 8), (7, 5)$
    (b) $(5\frac{1}{2}, 6\frac{1}{2})$ or $(5.5, 6.5)$

3 (a) $(7\frac{1}{2}, \frac{1}{2}), (7\frac{1}{2}, 4\frac{1}{2}), (9\frac{1}{2}, 4\frac{1}{2}), (9\frac{1}{2}, \frac{1}{2})$
    (b) $(8\frac{1}{2}, 2\frac{1}{2})$

4 (a) Inside A   (b) Inside C
    (c) Inside B   (d) Not inside any
    (e) Inside A   (f) Inside A

Sheet P60 (section C)   ▽□●

1 (a) $(1, 2), (1, ^-3), (^-4, ^-3), (^-4, 2)$
    (b) $(^-1\frac{1}{2}, ^-\frac{1}{2})$, or $(^-1.5, ^-0.5)$

2 (a) $(6, 2), (6, 7), (1, 7)$
    (b) $(3\frac{1}{2}, 4\frac{1}{2})$

3 (a) $(^-9, 2), (^-9, 7), (^-4, 7)$
    (b) $(^-6\frac{1}{2}, 4\frac{1}{2})$

4 $(^-6\frac{1}{2}, ^-5\frac{1}{2})$

Oral questions 1 (p 54)

The information table in the pupil's book is the basis for oral questions.

◊ Aim for regular sessions of oral questions. You can use the information table on several different occasions.

◊ Keep the questions simple when doing this work for the first time. Do not give too many questions: it is better to stop with all pupils feeling some sense of achievement after, say, five minutes rather than persist for half an hour!

◊ Start by explaining what is meant by the diameter of a ball, and that the diameter may vary slightly, which is why a range of values is given (for example, the men's shot can be any measurement between 11 and 13 cm in diameter). Similarly a range is given for the weight and cost in some cases. Encourage pupils to include units such as grams, centimetres and £.

◊ Some possible questions are given below. You can make up your own. You could also ask each pupil to write one question, and then give all the questions orally in class.

▼□○

1	What is the diameter of a golf ball?	4.3 cm
2	What is the diameter of a pool ball?	5.7 cm
3	What is the diameter of a tenpin bowl?	21.6 cm
4	How much is a rounders ball?	£4
5	How much is a golf ball?	£1.30
6	Which is larger – a table tennis or golf ball?	Golf ball
7	What is the weight of a golf ball?	45.9 g
8	What is the weight of a table tennis ball?	2.5 g
9	What is the weight of a tenpin bowl?	7258 g
10	What is the cost of a woman's shot?	£10
11	What is the cost of two men's shots?	£20
12	How much do three rounders balls cost?	£12

▽■○	1	How much is the cheapest football?	£7.50
	2	How much is the dearest football?	£30
	3	How much for a pair of tenpin bowls?	£80
	4	How much might you expect for one tenpin bowl? (Though it might not be possible to buy one!)	£40
	5	Which is the smallest ball?	Table tennis
	6	Which ball has the greatest diameter?	Football
	7	Which is the lightest ball?	Table tennis
	8	Which is the heaviest ball?	Men's shot
	9	Which balls could weigh 160 grams?	Cricket, pool
	10	Which balls could have a diameter of 5.7 cm?	Rounders, Pool
	11	What is the smallest diameter of a football?	21.8 cm
	12	What is the largest diameter of a rounders ball?	6.0 cm

▽□●	1	How much might one tennis ball cost?	£1.30
	2	How much might one table tennis ball cost?	50p
	3	How many rounders balls could you buy for £20?	5
	4	How many golf balls could you buy for £5?	3
	5	How many cricket balls could you buy for £50?	6
	6	How much for nine tennis balls?	£11.70
	7	How much for eight bowls and two jacks?	£224
	8	Give two balls that weigh roughly the same.	Men's shot, tenpin or cricket, pool or football, netball
	9	How much heavier is the men's shot than the women's shot?	3260 g
	10	How many cricket balls weigh just under 1 kg?	6

Extension You may like to extend this to converting units at a later date, for example:
What is the diameter in millimetres of a tennis ball?
What is the weight in kilos of a shot?

⑤ Brackets

The unit begins with a discussion of numerical expressions to help pupils see the need for brackets. It ends with 'Pam's game' which has been found successful in motivating pupils of all abilities. Algebraic expessions are not included.

You may have introduced brackets in earlier number work, for example S2 'Four digits'. This unit should provide consolidation and extension.

You may wish to discuss the convention that multiplication and division take priority over addition and subtraction in expressions such as $3 + 4 \times 7$ and $16 - 10 \div 2$. However, this could be confusing for some at this stage and no use is made of it in the unit.

p 55	**A** Check it out	▼■●	Using brackets to indicate which part of a calculation is to be done first Numerical expressions that use brackets
p 57	**B** Three in a row	▼■●	Practising using one set of brackets
p 57	**C** Brackets galore!	▽■●	Using more than one set of brackets
p 58	**D** Pam's game	▼■●	Practising using brackets

Essential

Dice

Practice sheets P61, P62

𝔸 **Check it out** (p 55)　　　　　　　　　　　　　　　▼■●

Teacher-led discussion introduces the idea that brackets indicate which part of a calculation is to be carried out first.

◊ Some of the expressions are correct as they stand, for example:
21 + 3, 30 − 6.
With some expressions, brackets can be used to indicate which part of the expression needs to be evaluated first, for example (Liz's second attempt):
$(3 + 3) \times 4 = 24$ but $3 + (3 \times 4) \neq 24$.
A few of the expressions are equivalent to 24 with brackets in any position, for example $(2 \times 6) \times 2$ and $5 + 3 + 20 - 4$. There is no need to labour this point at this stage.

One expression is incorrect: 13 + 21.

'We had already done "Four digits" and looked at brackets and calculators. This unit was a useful revision.'

◊ You can discuss how different calculators evaluate, say, $3 + 3 \times 4$, some working from left to right and others where multiplication and division take priority over addition and subtraction.

A5 Pupils can work in pairs or groups so solutions can be pooled and checked.

There are six different answers here but 24 calculations are possible (considering, for example, $(2 + 5) \times 3$ and $3 \times (2 + 5)$ to be different). High attainers could try to explain this. Pupils could choose three numbers and two operations to give more or fewer than six answers.

A6 Digits should not be joined to make larger numbers (24, 46 etc.).

Extension As an extension, pupils could investigate expressions that give the same value with brackets in any position. For example, $(2 + 3) - 5 = 2 + (3 - 5)$ and $(2 \times 8) \div 4 = 2 \times (8 \div 4)$.

They could try to find rules for when an expression is of this type.

B **Three in a row** (p 57)

This game consolidates the use of one set of brackets.

> Three dice for each pair/group (or one dice can be thrown three times), copies of the 'Three in a row' game board (pupils can copy this onto squared paper)

'Good fun, but weaker ones had to be surpervised closely.'

◊ Pupils play the game in pairs or in larger groups (split into two teams).

◊ The squares of the game board can be made large enough to use counters.

C **Brackets galore!** (p 57)

Pupils work with multiple and nested sets of brackets.

◊ The '4s make 7' activity below can be used as a homework.

> **4s make 7**
>
> Make up as many expressions as you can which have a value of 7.
>
> You can use any of $+, -, \times, \div$, and brackets and 4 as often as you like.
>
> Here are some examples to start you off.
> $(44 \div 4) - 4$
> $(4 + 4) - (4 \div 4)$
> $(4 + 4 + 4) \div 4) + 4$
> $(444 + 4) - (4 \times 4 \times 4) - 44 + 4$

Pam's game (p 58)   ▼■●

This game consolidates the use of brackets.

> One dice if game done as a class activity, otherwise enough dice for one
> for each group

◊ 'Pam's game' has been found successful in motivating pupils of all abilities
to use brackets. It works well as a class activity or in small groups.

In one school, a teacher split her whole class into two opposing teams and
set a time limit of 1 minute. Imposing a time limit may be necessary to
keep the game moving – some pupils always want to get 100 exactly!

◊ Some pupils (especially those who have completed section A only) may
adopt the convention that unless brackets show otherwise, an expression is
evaluated from left to right. For example, $6 \times (3 + 2) + (5 \times 4) \times 2$ may be
evaluated as 100 ($6 \times 5 + 20 \times 2$ worked from left to right). Some pupils
will appreciate the need for extra brackets here to give $((6 \times (3 + 2)) +
(5 \times 4)) \times 2$; others may not. It is likely you will want to emphasise this
point more or less strongly to different groups of pupils.

◊ You could play with fewer than six numbers and/or a different target.

Ⓐ **Check it out** (p 56)

A1 (a) 8   (b) 13   (c) 10   (d) 23
(e) 20   (f) 2   (g) 20   (h) 0
(i) 12   (j) 3   (k) 9   (l) 3

A2 (a) $(6 - 1) \times 3 = 15$
(b) $4 \times (1 + 2) = 12$
(c) $(2 + 1) \times 5 = 15$
(d) $(6 \div 3) + 9 = 11$
(e) $2 + (3 \times 4) = 14$
(f) $5 \times (2 - 1) = 5$
(g) $(5 - 1) \times 4 = 16$
(h) $2 + (2 \times 2) = 6$
(i) $3 \times (3 - 3) = 0$
(j) $(4 + 4) \div 4 = 2$
(k) $12 \div (3 \times 2) = 2$
(l) $10 - (6 - 2) = 6$

A3 A and Y, B and Z, C and X

A4 (a) $(1 + \mathbf{4}) \times 2 = 10$
(b) $(\mathbf{5} - 2) \times 4 = 12$
(c) $(3 \times \mathbf{3}) - 5 = 4$
(d) $2 \times (10 - \mathbf{8}) = 4$
(e) $4 \times (\mathbf{13} - 3) = 40$
(f) $(6 + \mathbf{9}) \div 3 = 5$
(g) $9 \div (\mathbf{5} + 4) = 1$
(h) $10 \div (6 - \mathbf{1}) = 2$
(i) $20 \div (\mathbf{4} + 6) = 2$
(j) $(4 \times 3) - (5 - \mathbf{2}) = 9$

A5 $(2 \times 3) + 5 = 11$
$(2 \times 5) + 3 = 13$
$(3 \times 5) + 2 = 17$
$2 \times (3 + 5) = 16$
$3 \times (2 + 5) = 21$
$5 \times (2 + 3) = 25$

There are only these six different
numbers, but 24 calculations are possible.

A6 Examples are

$6 - (4 - 2)$ $6 - (4 \div 2)$

$(6 - 4) \times 2$ $2 \times (6 - 4)$

$(6 - 4) + 2$ $2 + (6 - 4)$

$(6 + 2) - 4$

ℂ Brackets galore! (p 57)

C1 $(2 + 3) \times (10 - 3) = 35$

$((4 \times 20) \div 2) \times 5 = 200$

$(10 - (2 + 5)) \times 3 = 9$

$8 + (\, 2 \times (4 + 6)) = 28$

C2 (a)   There are four different possible values:

$(5 + 3) \times (4 - 1) = 24$

$((5 + 3) \times 4) - 1 = 31$

$(5 + (3 \times 4)) - 1 = 16$ or

$5 + ((3 \times 4) - 1) = 16$

$5 + (3 \times (4 - 1)) = 14$

(b)   There are five different possible values:

$(12 \div 2) + (4 \times 2) = 14$

$((12 \div 2) + 4) \times 2 = 20$

$(12 \div (2 + 4)) \times 2 = 4$

$12 \div ((2 + 4) \times 2) = 1$

$12 \div (2 + (4 \times 2)) = 1.2$

C3 There are six different possible values (one of which is negative):

$((9 \times 4) \div (2 \times 3)) - 1 = 5$ or

$(9 \times (4 \div (2 \times 3))) - 1 = 5$

$((9 \times 4) \div 2) \times (3 - 1) = 36$ or

$(9 \times (4 \div 2)) \times (3 - 1) = 36$ or

$9 \times ((4 \div 2) \times (3 - 1)) = 36$

$(9 \times 4) \div (2 \times (3 - 1)) = 9$ or

$9 \times (4 \div (2 \times (3 - 1))) = 9$

$(((9 \times 4) \div 2) \times 3) - 1 = 53$ or

$((9 \times (4 \div 2)) \times 3) - 1 = 53$ or

$(9 \times ((4 \div 2) \times 3)) - 1 = 53$ or

$(9 \times (4 \div 2) \times 3) - 1 = 53$

$(9 \times 4) \div ((2 \times 3) - 1) = 7.2$ or

$9 \times (4 \div ((2 \times 3) - 1)) = 7.2$

$9 \times ((4 \div (2 \times 3)) - 1) = {}^-3$

$9 \times (((4 \div 2) \times 3) - 1) = 45$

What progress have you made? (p 58)

1 (a)   15   (b)   10   (c)   10

    (d)   20   (e)   2   (f)   14

2 (a) $(6 + 1) \times 2 = 14$

    (b) $3 \times (10 - 8) = 6$

    (c) $10 - (6 \div 2) = 7$

Practice sheets

Sheet P61 (section A)

1 (a)   11   (b)   16   (c)   8

    (d)   3   (e)   4   (f)   30

    (g)   6   (h)   4   (i)   12

2 (a) $4 \times (3 + 2) = 20$

    (b) $(4 \times 3) + 2 = 14$

    (c) $10 - (3 \times 2) = 4$

    (d) $(10 - 3) \times 2 = 14$

    (e) $(4 + 5) \times 2 = 18$

    (f) $4 + (5 \times 2) = 14$

    (g) $(12 \div 3) + 1 = 5$

    (h) $12 \div (3 + 1) = 3$

    (i) $9 + (4 \div 2) = 11$

3   A and X   B and Y

    C and V   D and Z

4 (a) $(2 \times \mathbf{3}) + 1 = 7$

(b) $(3 + \mathbf{2}) \times 2 = 10$

(c) $\mathbf{3} + (4 \times 2) = 11$

(d) $9 - (\mathbf{3} + 2) = 4$

(e) $(\mathbf{6} \div 2) + 5 = 8$

(f) $12 \div (1 + \mathbf{5}) = 2$

(g) $2 \times (8 - \mathbf{5}) = 6$

(h) $10 - (\mathbf{7} - 3) = 6$

5 There are five different possible numbers:

$4 + (2 \times 3) = 10$ (the example)

$(4 + 2) \times 3 = 18$

$2 + (3 \times 4) = 14$

$(2 + 3) \times 4 = 20$

$3 + (4 \times 2) = 11$

These results can be found in different ways, for example $(4 + 3) \times 2 = 14$

Sheet P62 (section C)

1 (a) $(2 \times 6) - (3 + 1) = 8$

(b) $2 \times (6 - (3 + 1)) = 4$

2 (a) $(12 - 5) - (2 - 1) = 6$

(b) $((12 - 5) - 2) - 1 = 4$

3 (a) 4   (b) 10   (c) 0

(d) 6   (e) 10

4 $((15 - 6) - 4) - 1 = 4$

$(15 - (6 - 4)) - 1 = 12$

$(15 - 6) - (4 - 1) = 6$

$15 - ((6 - 4) - 1) = 14$

$15 - (6 - (4 - 1)) = 12$

5 $((5 + 2) \times 3) + 4 = 25$

$(5 + (2 \times 3)) + 4 = 15$

$(5 + 2) \times (3 + 4) = 49$

$5 + ((2 \times 3) + 4) = 50$

$5 + (2 \times (3 + 4)) = 19$

6 (a) $(12 + 6) \div (3 - 1) = 9$

(b) $(12 + (6 \div 3)) - 1 = 13$

(c) $((12 + 6) \div 3) - 1 = 5$

(d) $12 + (6 \div (3 - 1)) = 15$

Oral questions 2 (p 59)

'It was the last 15 minutes of the lesson and lunch time was approaching. Most pupils did very well making only one or two small mistakes or getting all correct. It was very good diagnostically.'

◊ The information on the pupil's page is intended for use as a context for orally given questions. Some sample questions are given after these notes.

◊ Oral work of this kind is usually most successful when pupils work in silence. Explain that you will read each question twice and then they will be given time to answer it. Some schools allowed 15 seconds between questions, others allowed more. Emphasise that pupils need to listen very carefully and if they cannot do a question they should forget it and try the next one.

You may like to record the questions on a tape with a specified time limit between questions and play that in class. This can encourage pupils to concentrate on listening to the questions.

◊ An initial discussion of pupils' own calculating methods may be beneficial, especially ways of calculating mentally with money.

◊ As a follow-up activity, pupils can invent their own questions for the rest of the class. In one school, each pupil wrote their own question on a slip of paper and put the answer on the back. The questions were collected together and read out for the whole class to answer.

▼□○
1	Find four items which cost under £1 each.	Note pad, pencils, ruler, eraser
2	What is the total cost of a note pad and an eraser?	£0.72 or 72p
3	How much change from a £10 note do I get if I buy the stapler?	£2.20
4	Find the cost of four packs of ink cartridges.	£4.28
5	A Stick-it note measures 4 cm by 5 cm. Draw it accurately.	
6	Find the cost of a pencil case and an eraser.	£1.25
7	Pencil cases go up in price by 10p. How much does one cost now?	£1.20
8	What is the most expensive item?	Hole punch
9	Which item costs £0.79?	Ruler

▽■○	**1**	Find the total cost of the two most expensive items.	£16.30
	2	How much is a ruler and a note pad?	£1.36
	3	Find the cost of four note pads.	£2.28
	4	Find the cost of ten rulers.	£7.90
	5	I buy the sticky tape and the scissors. How much is this?	£4.34
	6	The shop has a half-price sale. What is the sale price of the hole punch?	£4.25
	7	What is the cost of two rulers?	£1.58
	8	I buy a hole punch and a pencil case. How much change will I get from £10?	£0.40 or 40p
	9	How much would the stapler cost if the price was reduced by 50p?	£7.30
	10	What is the cost of two geometry sets?	£10.66
▽□●	**1**	What is the difference in cost between the dearest and cheapest article?	£8.35
	2	How much will pencils for a class of 30 cost?	£3.35
	3	How many erasers could you buy for £1?	6
	4	How much is a geometry set and a ruler?	£6.12
	5	How much for the stapler and colouring pencils?	£11.00
	6	One Stick-it note has a width of 4 cm and an area of 20 cm^2. What is the length of the note?	5 cm
	7	How many rulers can you buy for £3.00?	3
	8	What is the total cost of the geometry set, the hole punch and the stapler?	£21.63
	9	How much do you think one pencil should be sold for?	11p or 12p
	10	The shop has a half-price sale. What is the sale price of the colouring pencils?	£1.60
	11	There are five maths teachers in a school. How much would it cost to buy each teacher a hole punch?	£42.50
	12	How many packs of ink cartridges could you buy with £12 and how much money would you have left?	11, 23p left

Review 1 (p 60)

1 (a) (b)

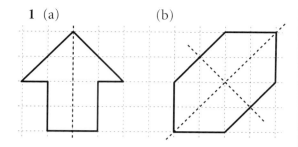

2 (a) 17 (b) 10 (c) 14 (d) 20

3 A (7, 2) B (4, 4) C (0, 2) D (3, 0)

4 (a) $2 \times (5 + 3) = 16$
  (b) $(20 - 10) \div 2 = 5$
  (c) $20 - (10 \div 2) = 15$

5 (a) Line 2   (b) Lines 1 and 2

6 (a) $(5 + \mathbf{3}) \div 2 = 4$
  (b) $12 - (\mathbf{8} + 1) = 3$
  (c) $6 \times (\mathbf{6} - 1) = 30$

7 (a) (7, 3)   (b) (2, 3)
  (c) (3, 6)   (d) (4, 4)

8 (a)

  (b)

9 (a) (⁻4, ⁻2)   (b) (⁻7, 1)
  (c) (⁻6, ⁻1) (⁻1, ⁻6)

10 $(2 + 4) \div 8 = (4 + 2) \div 8 = 0.75$
  $2 + (4 \div 8) = (4 \div 8) + 2 =$
  $(2 + 8) \div 4 = (8 + 2) \div 4 = 2.5$
  $(2 \div 4) + 8 = 8 + (2 \div 4) = 8.5$
  $2 \div (4 + 8) = 2 \div (8 + 4) = 0.1666...$
  $2 + (8 \div 4) = (8 \div 4) + 2 = 4$
  $(2 \div 8) + 4 = 4 + (2 \div 8) = 4.25$
  $(4 \div 2) + 8 = 8 + (4 \div 2) = 10$
  $4 \div (2 + 8) = 4 \div (8 + 2) = 0.4$
  $(4 + 8) \div 2 = (8 + 4) \div 2 = 6$
  $4 + (8 \div 2) = (8 \div 2) + 4 = 8$
  $8 \div (2 + 4) = 8 \div (4 + 2) = 1.333...$

Growing patterns

Pupils investigate sequences arising from a variety of contexts.

The emphasis is on finding a rule to continue a sequence and explaining why the rule is valid. There is a little work on finding a rule that effectively gives the nth term (though not expressed in that way), but this aspect is covered more fully in Unit 12 'Work to rule'.

Pupils should begin to realise that just spotting a pattern in the first few numbers in a sequence (for example, 'add 3 to the previous number') is not enough to prove that the sequence will continue in the same way. You have to go back to the context (rose bushes, ponds, earrings, ...) to give a convincing explanation.

In expressions such as $(2 \times$ *number of red rose bushes*$) + 2$, brackets are used. You may already have introduced the convention that multiplication and division take priority over addition and subtraction, in which case the brackets can be dropped.

T	p 62 **A** Coming up roses	▼■●	Investigation leading to a linear sequence Describing how the sequence continues Explaining why the sequence continues like this
	p 62 **B** Pond life	▼■●	Investigation leading to a linear sequence
	p 64 **C** Changing shape	▼■●	Consolidating work on linear sequences
T	p 65 **D** Earrings	▼■●	Investigation leading to a simple exponential sequence (powers of 2)
	p 66 **E** Staircases	▽■●	Investigation leading to a Fibonacci sequence

Optional
Coloured counters
Coloured tiles
Multilink cubes
Square dotty paper
Sheet 70
Sheets 71 and 72 (▽■●)

Practice sheets P63 to P65

This investigation gives rise to a linear sequence.
Pupils describe and explain how the sequence continues.

> Optional: Tiles/counters to represent red and white rose bushes

◊ A gardener is designing a display with red and white rose bushes planted as single rows of red roses, with a row of white roses on each side and a white rose at each end. Some examples are:

    🌹 represents a white rose bush.

    🌹 represents a red rose bush.

◊ To start with, pupils could think about the designs on the page and then try to draw similar arrangements that use 2, 6, 4 red roses for example.

Encourage pupils to simplify the diagrams, for example by using coloured circles or the letters R and W.

Pupils count the number of red and white rose bushes needed in each arrangement so far and collect their results together in an ordered table. Discuss the advantages of tabulating in this way.

Alternatively, pupils could consider what the smallest design would look like and draw it. A sequence of designs can then be produced in order: 1 red rose bush, 2 red rose bushes etc. These results can then be tabulated.

◊ Ask pupils to complete the table up to, say, 8 red rose bushes.
Discuss any methods that pupils use to complete the table.
These may include

• making/drawing the designs

• finding and using the rule that the number of white rose bushes increases by 2 for every extra red rose bush

• finding and using the rule that the number of white rose bushes is (2 × *the number of red rose bushes*) + 2

It is important that each method is considered equally valid.

◊ In discussion, bring out the fact that the number of white rose bushes increases by 2 for every extra red rose bush. Ask the pupils to use the diagrams to explain why this is so.

'Section A worked very well as an introduction to how to tackle an investigation generally. We spent quite a bit of time on it which meant that sections B and C went well with pupils tackling the questions in a systematic way. It also meant that we didn't have time to do section E!'

You could ask the pupils to consider a range of 'explanations'.
For example, the number of white rose bushes increases by 2 for every extra red rose bush because

- the numbers in the table go up in 2s
- the row of red bushes has 1 white bush at each end (2 in total)
- there are 2 colours of roses
- each red rose bush has 1 white rose bush either side of it (2 in total)

These statements could be put on cards and groups of pupils could consider which is an acceptable explanation.

Many pupils would think that just describing how the sequence continues (as in the first statement above) is a perfectly acceptable 'explanation'. Emphasise they must refer back to the arrangement of bushes to explain why the sequence will continue in the same way.

◊ Now ask pupils to think about a larger number of red bushes, for example 20 red bushes, and to say how many white bushes would be needed for them. Again, pupils are likely to use a variety of methods as before.

If pupils use the rule that the number of white rose bushes is (2 × the number of red rose bushes) + 2, ask them to explain why they know their rule works by referring to the arrangement of rose bushes. Emphasise that just because their rule works for a few results it does not follow it will work for all results.

Pose a question like '314 white rose bushes are needed for 156 red rose bushes. So how many will be needed for 157 rose bushes?' and ask pupils to discuss their methods. High attainers can discuss the most appropriate method for each of the two types of problem.

B Pond life (p 62)

Pupils investigate another situation that gives rise to a linear sequence. They describe and explain how the sequence continues.

> Optional: Square dotty paper, square tiles

◊ Make sure pupils are clear that the ponds are square ponds.

B5 A variety of methods are possible. If pupils are struggling ask them to look at their answer for B4(d) and think how that could help them. Some pupils will continue to draw ponds at this stage.

B6 In part (b), some pupils may say 'because the numbers in the table go up in 4s'. Emphasise that they must refer back to the arrangement of slabs to explain why they can be sure that the sequence continues in the same way.

B7 Ask those who use the 'multiply by 4 and add 4' rule or the 'add 1 and multiply by 4' rule to consider if there is any advantage in using the 'going up in 4s' rule here (the calculation $272 + 4$ is simpler than $(68 \times 4) + 4$).

B8 Some pupils will solve this problem by counting on in 4s. Encourage higher attainers to think of a more direct method. They can then compare the best ways to solve B7 and B8.

C **Changing shape** (p 64)   ▼■●

This consolidates work on linear sequences.

> Optional:
> Square dotty paper, triangular dotty paper
> Square tiles

C1 In part (b), emphasise that the width of all the ponds for this table is 3 metres. The length of the pond is the other dimension so, for some ponds, the length is shorter than the width.

C5 All pupils could try this but some will need help to structure their work.

Extension Pupils could consider triangular ponds surrounded by triangular slabs.

D **Earrings** (p 65)   ▼■●

Pupils investigate a situation leading to a simple exponential sequence (powers of 2). They find and explain how the sequence continues.

> Optional:
> Red and yellow multilink cubes
> Sheet 70 (for recording results)

◊ Multilink has been found to be a very useful way to 'build' the earrings. It allows easy identification of duplicates and collection of results.

If pupils use multilink, make sure they realise that these two designs are different.

◊ Ask pupils to find as many different three-bead earrings as they can.

Collect the results for the whole class and discuss how they can be sure that they have found all possible designs for three beads.

The 8 different designs are:

◊ It is important to stress to high attainers that 'predict and check' is a good strategy to increase confidence that any patterns or rules found are correct. However, predict and check does not explain **why** any patterns or rules are valid and hence is not a proof that the sequence of numbers will continue in the way they expect.

D3 In part (c), pupils can record their results on sheet 70.

D4 Pupils cannot claim to be sure about the number of five-bead earrings until they have found them all (and shown no more exist) or until they have explained why the numbers in the sequence double each time.

D5 In part (b), most pupils will find it difficult to explain why the number of earrings doubles each time. If so, encourage them to look at their sets of earrings for say 3 beads and 4 beads and to consider how they are related.

Extension If the number of ways of using 0 reds, 1 red, 2 reds, … is considered for each length of earring, Pascal's triangle arises.

E **Staircases** (p 66)   ▽■●

Pupils investigate a situation that gives rise to a Fibonacci sequence.

> Optional: Sheets 71 and 72 (for recording results)

◊ In some schools, this activity worked with the whole range of attainment.

◊ Emphasise that the staircase in the diagram has four steps and not five. This can confuse pupils and using sheets 71 and 72 may help.

◊ Ask pupils to think carefully about ways of recording their results. Pupils who choose to record their results as sequences of 1s and 2s may realise that the problem reduces to that of finding how many different ways a total can be reached by adding 1s and 2s.

E3 Cover up the last two entries in the table and ask pupils to imagine they were trying to predict the number of ways to climb four steps. It's very tempting to predict 4 ways, the sequence 1, 2, 3, … increasing by 1 each time. However, the prediction would be incorrect. This is an opportunity to reinforce the dangers of relying on an apparent number pattern.

E6 This is extremely difficult for most pupils but should provide extension for the highest attainers. A possible explanation is included in the answers.

Extension Extend the investigation so that three steps can be climbed at a time.

The results for staircases up to six steps are shown in the table below.

Number of steps	1	2	3	4	5	6
Number of different ways	1	2	4	7	13	24

Each number of ways is found by adding the previous three.

B Pond life (p 62)

B1 (a) 8 slabs (b) 12 slabs (c) 16 slabs

B2 (a)   (b) 20 slabs

B3 (a) 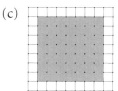 (b) 24 slabs

B4 (a) The numbers of slabs in the table are 8, 12, 16, 20, 24, 28.

(b) 32 slabs (c)

(d) 44 slabs
Pupils' methods are likely to involve
- counting on in 4s or
- multiplying by 4 and adding 4 or
- adding 1 and multiplying by 4

B5 11 metres
Pupils' methods could involve
- extending the pattern in the table or
- working from the fact that a 10 by 10 pond needs 44 slabs or
- subtracting 4 and dividing by 4 or
- dividing by 4 and subtracting 1

B6 (a) The number of slabs needed goes up by 4 each time.

(b) The pupil's explanation: for example, an increase of 1 metre in width means an extra slab for each edge. Since the pond has 4 edges, 4 extra slabs are needed.

B7 276 slabs

B8 64 slabs
The pupil's method: for example, $(15 \times 4) + 4 = 64$

B9 B $(n \times 4) + 4$

C Changing shape (p 64)

C1 (a) (i) 20 slabs (ii) 14 slabs

(b) The numbers of slabs in the table are 12, 14, 16, 18, 20, 22, 24.

(c) 26 slabs
The pupil's method: for example, $24 + 2 = 26$

C2 (a) The number of slabs needed goes up by 2 each time.

(b) The pupil's explanation

C3 50 slabs
The pupil's method: for example, $(20 \times 2) + 10 = 50$

C4 C $(x \times 2) + 10$

C5 The pupil's investigations

D Earrings (p 65)

D1 4 different earrings

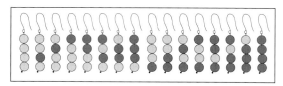

D2 The numbers of different earrings in the table are 2, 4, 8.

D3 (a) 16 different earrings

(b) The pupil's method: for example, $8 \times 2 = 16$ or $8 + 8 = 16$

(c) One way to organise the results is to add a yellow bead to each of the three-bead earrings and then add a red bead.

Another way is to look at earrings with 0 red beads, 1 red bead, 2 red beads, …

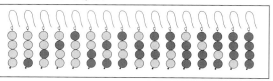

D4 32 different earrings

D5 (a) The number of earrings doubles each time.

    (b) The pupil's explanation

E **Staircases** (p 66)

E1 5 ways

E2 (a) 8 ways

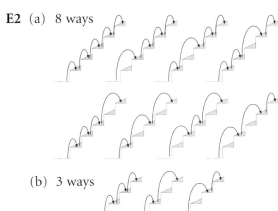

    (b) 3 ways

E3 The numbers of different ways in the table are 1, 2, 3, 5, 8.

E4 (a) 13 ways

    (b) The pupil's method: for example, add the previous two numbers in the sequence, 5 + 8 = 13.

    (c)

E5 To find the number of ways to climb a staircase, add the number of ways to climb the previous two staircases.

E6 The pupil's explanations. A possible explanation is as follows: Consider for example a staircase with 5 steps. Each set of 'steps' up the staircase ends in a '1-step' or a '2-step'. Find the ways that end in a '1-step' by considering all the ways to climb 4 steps (add a '1-step' on to each of them). Find the ways that end in a '2-step' by considering all the ways to climb 3 steps (add a '2-step' on to each of them). So find the total number of ways to climb 5 steps by adding the number of ways to climb 4 steps to the number of ways to climb 3 steps. A similar argument works for any size staircase.

What progress have you made? (p 67)

1 (a) 6 bushes   (b) 8 bushes

2

3 The numbers of yellow rose bushes in the table are 5, 6, 7, 8, 9.

4 (a) 12 yellow bushes

    (b) The pupil's method: for example, count in 1s from the result for 5 red bushes.

5 (a) The number of yellow bushes goes up by 1 each time.

    (b) The pupil's explanation: for example, an increase of 1 red bush means an extra yellow bush in the bottom row.

6 (a) 104 yellow bushes

    (b) The pupil's method: for example, 100 + 4 = 104

7 (a) 46 red bushes

    (b) The pupil's method: for example, 50 − 4 = 46

Practice sheets

Sheet P63 (sections A, B, C) ▼■●

1 The pupil's check

2 7 corners

3 The pupil's chain of 4 squares; 13 corners

4 The numbers of corners in the table are 4, 7, 10, 13.

5 (a) 16 corners

   (b) The pupil's drawing of a 5-square chain

6 31 corners

7 (a) The number of corners goes up by 3 for each added square.

   (b) The corner of a new square that joins the chain has already been counted, so 3 corners are added each time.

8 61 corners

9 C $(n \times 3) + 1$

Sheet P64 (section D) ▼■●

1 The pupil's drawing for year 4; 8 flowers

2 The numbers of flowers in the table are 1, 2, 4, 8.

3 16 flowers

4 The pupil's drawing for year 5, with 16 flowers

5 The number of flowers doubles each year.

6 Each year every flower from the year before is replaced by two flowers.

Sheet P65 (section E)

1 (a) 5 ways   (b) 8 ways

2 The numbers of different ways in the table are 1, 2, 3, 5, 8.

3 (a) 13 ways

   (b) The pupil's method: for example, 5 + 8 = 13

   (c) 1, 1, 1, 1, 1, 1   1, 1, 2, 2
        1, 1, 1, 1, 2   1, 2, 1, 2
        1, 1, 1, 2, 1   2, 1, 1, 2
        1, 1, 2, 1, 1   1, 2, 2, 1
        1, 2, 1, 1, 1   2, 1, 2, 1
        2, 1, 1, 1, 1   2, 2, 1, 1
                                 2, 2, 2

4 (a) To find the number of different ways, add together the number of ways for the previous two values.

   (b) A possible explanation is as follows: Consider for example stamps that give a value of 5p. Each row of stamps ends in a 1p or a 2p stamp. Find the number of rows that end in a 1p stamp by considering all the ways that give 4p (add a 1p stamp on to the right of each of them). Find the number of rows that end in a 2p stamp by considering all the ways that give 3p (add a 2p stamp on to the right of each of them). So find the total number of ways that give 5p by adding the number of ways for 4p to the number of ways for 3p. A similar argument works for any value.

   Some pupils may see that the 'stamps' problem is in essence the same as the 'staircase' problem.

8 Angle

T	p 68	**A** Making angles	▼■●	Making the connection between angles and turning
	p 69	**B** Comparing angles	▼□○	Comparing the sizes of angles by tracing and by eye
	p 70	**C** Right angles, acute, obtuse and reflex angles	▼■○	
T	p 72	**D** Measuring angles	▼■●	
T	p 74	**E** Drawing angles	▽■●	
	p 75	**F** Angles round a point	▽■●	
T	p 76	**G** Calculating angles	▽■●	Using angles round a point, angles on a line, vertically opposite angles
	p 78	**H** Find it!	▽□●	Following instructions about angles and distances to locate points on a map

Essential

Sheet 73 (for you: one on white card, one on coloured card)
Sheet 74 (for each pupil: one circle on white card, one on coloured card)
Sheet 75 or 76 or 77 (see below, section D)
Sheet 78 (▽□●)
Tracing paper
Angle measurers

Practice sheets P66, P67, P68 and P69

T

A Making angles (p 68) ▼■●

Use sheet 73 to make a large angle-maker for yourself.
Each pupil needs a small angle-maker, made from:

- one circle from sheet 74 on coloured card
- one circle from sheet 74 on white card

If pupils are to make their own, it is better if they follow a demonstration from you. (You can keep them for other groups.)

◊ You can use the scissors pictures to find out what pupils know already about angle. Do they, for example, think that the size of the scissors affects the size of the angle?

Using the angle-maker

◊ Turn the coloured circle gradually to make each quarter turn like this.

  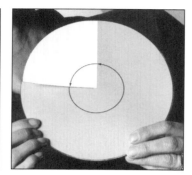

Ask pupils to describe each of the coloured angles in as many ways as they can. Record the words they use and link them together. For example, 90° = right angle = quarter turn.

Set the angle-maker to zero again, and start with the line neither vertical nor horizontal. Ask pupils to tell you to stop turning when you have made

- a right angle
- a half turn
- three quarters of a turn

Some pupils find it difficult to recognise these angles in different orientations.

◊ Introduce the terms acute, obtuse and reflex.

You could define them in terms of quarter turn, half turn, etc. (unless pupils are already familiar with degrees). Make some more angles in various orientations for pupils to describe as acute, obtuse or reflex.

Comparing angles

◊ Some pupils see this as a 'large' angle:

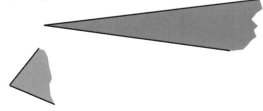

and this as a 'small' angle:

The purpose of the following work is to emphasise that the angle is the amount of turn and has nothing to do with the lengths of the arms or the area between them.

'I did do this and it worked very well. I didn't spend too much time on it as almost all of the class knew quite a bit about angles already.'

Show an angle on your angle-maker. Ask pupils to make it on their smaller ones. Invite some of them to compare their angle to yours by placing it over yours. Emphasise that the lengths of the arms doesn't matter.

Make an angle on a large angle-maker and another on a small angle-maker. Ask which is the bigger angle. Include examples where

- the larger angle is on the smaller angle-maker
- both angles are the same
- the larger angle is on the larger angle-maker
- the two angles are held in different orientations

Describing angles

◊ Pupils work in pairs, sitting back to back. One makes an angle and describes it as closely as they can. The other makes the angle from the description. Then they compare the angles and see how close they were.

B **Comparing angles** (p 69) ▼□○

Tracing paper

This work can help you identify pupils who are still unsure about what makes one angle bigger than another.

C **Right angles, acute, obtuse and reflex angles** (p 70) ▼■○

D **Measuring angles** (p 72) ▼■●

Angle measurers (360°), sheets 75, 76, 77 (graded in difficulty; see D1 below)

◊ Use an OHP to demonstrate how to use the angle measurer. Show that you can measure starting from either arm. Discuss which scale to use. Link degrees to fractions of a turn.

◊ Draw some angles. Ask pupils to estimate then measure each of them.

D1 The three sheets are graded as follows:
- sheet 75   angles are all multiples of 5°
- sheet 76   angles are of any size
- sheet 77   angles are marked in the corners of polygons

Tilting bus

This is for stimulating discussion. The pointer on the body shows how far the body has tilted. The other pointer shows how far the chassis has tilted. The difference is due to the 'give' in the suspension system.

A double decker bus must remain stable when the chassis is tilted to 28° (in practice it can go much further than this). The test has to be done with weights added to simulate a full load of passengers upstairs but no extra weight downstairs.

E Drawing angles (p 74)

Angle measurers

Some pupils are reluctant to extend lines beyond the point they have marked against the angle measurer.

F Angles round a point (p 75)

Compasses

G Calculating angles (p 76)

These calculations use angles round a point, angles on a line and vertically opposite angles.

High attainers can make up their own questions similar to G5 and give them to others to do.

H Find it! (p 78)

Pupils follow instructions about angles and distances to locate points on a map.

Sheet 78

Make sure that pupils understand about turning through an angle from the direction in which they are initially facing. You could demonstrate this by walking forward holding a ruler out in front of you to show the direction in which you are facing. When you stop and turn, you move the ruler through an angle.

Now draw your path and the angle of turn on the board (as in the diagram in the pupil's book).

Pupils with experience of LOGO may be familiar with this use of angles.

B1 Angles *a* and *c* are bigger than *X*.
Angles *b* and *d* are smaller than *X*.

B2 (a) Bigger   (b) Bigger   (c) Smaller
     (d) Bigger   (e) Bigger

B3 *c, a, b, d*

C **Right angles, acute, obtuse and reflex angles** (p 70)

C1 *a* and *c* are right angles.

C2 (a) Acute   (b) Right angle
     (c) Acute   (d) Obtuse
     (e) Obtuse

C3 *a* is a right angle,   *b* is acute,
     c is acute,   *d* is obtuse,
     e is acute,   *f* is acute.

C4 *a* is acute,   *b* is obtuse,
     c is acute,   *d* is reflex,
     e is obtuse,   *f* is a right angle.

C5 *a* is acute,   *b* is a right angle,
     c is reflex,   *d* is a right angle,
     e is obtuse,   *f* is reflex,
     g is acute.

D **Measuring angles** (p 72)

D1 Pupils' answers should be within 1° of the answers given.

Sheet 75

$a = 30°$, acute $b = 40°$, acute
$c = 55°$, acute $d = 135°$, obtuse
$e = 15°$, acute $f = 265°$, reflex
$g = 90°$, right angle $h = 235°$, reflex

Sheet 76

$a = 15°$, acute $b = 90°$, right angle
$c = 50°$, acute $d = 93°$, obtuse

$e = 123°$, obtuse $f = 212°$, reflex
$g = 270°$, reflex $h = 117°$, reflex

Sheet 77

$a = 70°$, acute $b = 60°$, acute
$c = 107°$, obtuse $d = 73°$, acute
$e = 107°$, obtuse $f = 47°$, acute
$g = 112°$, obtuse $h = 56°$, acute
$i = 60°$, acute $j = 60°$, acute

D2 *a* is equal to *e*; *b* is equal to *f*.

D3 (a) $x = 33°$, $y = 28°$, $z = 119°$
     (b) They add up to 180°.
     (c) You should get the same total for any triangle.

D4 (a) $a = 100°$, $b = 139°$, $c = 48°$, $d = 73°$.
     (b) They add up to 360°.
     (c) You should get the same total for any four-sided shape.

E **Drawing angles** (p 74)

E1 90°

E2 50°

E3 40°

E4 (a) 80°   (b) 360°

F **Angles round a point** (p 75)

F1 30°

F2 (a) 60°   (b) 90°   (c) 150°

F3 270°

F4 (a) 150°   (b) 210°   (c) 270°
     (d) 180°   (e) 360°

F5 (a) 360°   (b) 72°
     (c) The pupil's drawing of a regular pentagon
     (d) The pupil's check

F6 The angles at the centre are
45° for a regular octagon
40° for a regular nonagon
36° for a regular decagon
30° for a regular dodecagon

Ⓖ Calculating angles (p 76)

G1 $a = 80°$ $b = 145°$ $c = 60°$ $d = 250°$

G2 $a = 140°$ $b = 35°$ $c = 130°$ $d = 115°$

G3 $a = 39°$ $b = 43°$ $c = 97°$
    $d = 234°$ $e = 151°$

G4 $a = 110°$ $b = 70°$ $c = 110°$
    $d = 17°$ $e = 163°$ $f = 17°$
    $g = 62°$ $h = 25°$ $i = 93°$ $j = 25°$

G5 $a = 102°$ $b = 36°$ $c = 123°$ $d = 36°$
    $e = 88°$ $f = 129°$ $g = 113°$ $h = 134°$
    $i = 22°$ $j = 125°$ $k = 125°$

Ⓗ Find it! (p 78)

Sheet 78
The spy camera is at (11, 14).
The X-ray binoculars are at (13, 9).
The coding kit is at (5, 14).
The radio is at (24, 13).

What progress have you made? (p 79)

1 a is an acute angle, b is a right angle,
c is an acute angle, d is an obtuse angle.

2 a, c, b, d

3 $a = 49°$ $b = 90°$ $c = 84°$ $d = 136°$

4 80°

5 $a = 47°$ $b = 110°$ $c = 70°$ $d = 101°$

Practice sheets

Sheet P66 (section A, B, C)   ▼■●

1 b, a, d, c

2 p is obtuse, q is a right angle,
   r is acute, s is acute.

3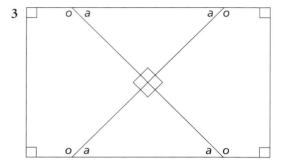

Sheet P67 (section D)   ▼■●

$a = 32°$ $b = 61°$ $c = 128°$
$d = 90°$ $e = 14°$

Sheet P68 (section F)   ▽■●

1 (a) NE   (b) W   (c) W

2 (a) NW   (b) S   (c) NE

3 135°

4 (a) 90°   (b) 270°   (c) 180°
   (d) 135°   (e) 225°   (f) 315°

5 (a) 120°   (b) 150°

6 (a) 210°   (b) 330°

Sheet P69 (section G)   ▽■●

$a = 170°$ $b = 60°$ $c = 45°$
$d = 36°$ $e = 130°$ $f = 105°$
$g = 64°$ $h = 64°$ $i = 116°$
$j = 60°$ $k = 25°$ $l = 65°$
$m = 34°$ $n = 80°$ $o = 140°$
$p = 96°$ $q = 75°$ $r = 72°$
$s = 75°$ $t = 33°$

9 Balancing

This work introduces, in an informal way, the idea of doing the same thing to both sides of an equation.

> **Essential**
> OHP
> Transparency copied from sheet 79
>
> **Practice sheets** P70 to P73

Ⓐ **Scales** (p 80)   ▼■●

This section introduces the idea of scales balancing. The emphasis is on what can be done to both sides and still leave the scales in balance.

> An OHP, a transparency of sheet 79, cut up so that each picture is separate.

◊ **Picture 1**

Use the OHP pictures of weights and rabbits to discuss with the class what you can **add** to both sides of the scales and still keep them in balance. Point out that all the weights are (notionally) 1 kilogram each.

T

'This was excellent with an OHP.'

'I didn't have an OHP, so I did my own silly drawings on the whiteboard.'

You could also ask the pupils what you can do to the scales and definitely make them unbalanced.

Picture 2

You can then use picture 2 to discuss how to find out what one hedgehog weighs, by taking things from both sides of the scales.

B Balance pictures (p 80) ▼□○

Pupils are not expected to show any working when solving these puzzles. Of course, if they wish to write down any intermediate steps they should not be discouraged from doing so.

C Writing (p 84) ▽■●

> Optional: OHP and the cut-up transparencies used for section A.

◊ When introducing this section to pupils, point out that we are simply recording what is done to the balance in the most straightforward way possible. You could use the cut-up pictures and the OHP when doing this.

Emphasise that the unknown here (n in the example) stands for the *number of kilograms* each animal weighs. Avoid using letters like h (which pupils may think stands for the word 'hedgehog') or w (which pupils may think of as standing for 'a weight').

C3 Pupils choose their own letters to stand for the weight of each animal. Emphasise that the letter stands for the weight of the animal, not the name of the animal or the animal itself.

D Weights and objects on both sides (p 86) ▽■○

◊ In this section, pupils continue to use the notation $n + n + n$ when solving problems. Section E introduces the shorthand $3n$ for $n + n + n$. Some pupils may themselves suggest using this shorthand, but do not force pupils into using the algebraic shorthand before they are ready for it.

This is the first time that pupils have to take both weights and unknown objects from each side. The emphasis is on writing out the solution to the puzzle step by step.

E Using shorthand (p 88) ▽□●

> 'We were all relieved when the shorthand was introduced!'

This section encourages higher attainers to use the notation $3n$ as a shorthand for $n + n + n$ when solving problems.

ℬ Balance pictures (p 80)

B1 (a) 4   (b) 3

B2 (a) 6   (b) 2   (c) 4
(d) 5   (e) 7   (f) 12
(g) 8   (h) 4   (i) 1
(j) 3

B3 5

B4 (a) 20   (b) 3   (c) 6
(d) 8   (e) 3   (f) 5

B5 5

B6 (a) 6   (b) 2   (c) 5   (d) 3

B7 2

B8 (a) 3   (b) 4   (c) 4   (d) 5

B9 The pupil's balance puzzle

B10 The pupil's puzzle and explanation of how to solve it

ℂ Writing (p 84)

In this and the following sections, pupils should write down their working and check.

C1 $n = 4$

C2 $x = 3$

C3 (a) 5   (b) 4   (c) 11   (d) 20

𝔻 Weights and objects on both sides (p 86)

D1 $n = 2$ and check

D2 (a) 4   (b) 3   (c) 2

D3 (a) 3   (b) 2

D4 The pupil's picture for
$s + s + s + 6 = s + 12$, with $s = 3$

D5 (a) $y = 12$ (b) $p = 3$ (c) $z = 2$
(d) $g = 2\frac{1}{2}$ (e) $n = 6$ (f) $t = 3$
(g) $w = 2$ (h) $d = 2\frac{1}{2}$ (i) $b = 20$
(j) $k = 5$

D6 The pupil's puzzle for a friend to solve

𝔼 Using shorthand (p 88)

E1 The pupil's check that $n = 2$ fits the puzzle

E2 (a) 4   (b) 3   (c) 2

E3 (a) 8   (b) 79

E4 The pupil's picture for $3h + 15 = h + 37$, solution $h = 11$

E5 (a) $p = 7$ (b) $d = 6$ (c) $s = 8$
(d) $q = 10$ (e) $n = 4$ (f) $u = 53$
(g) $k = 2\frac{1}{2}$ (h) $y = \frac{1}{3}$ (i) $a = 5$
(j) $w = \frac{1}{2}$ (k) $b = 20$ (l) $f = 3.2$
(m) $j = 17$ (n) $s = 2$

E6 (a) $d = 10$ (b) $e = 6$ (c) $k = 0.9$
(d) $y = 3.5$

E7 The pupil's puzzle for a friend to solve

What progress have you made? (p 90)

1 4

2 The pupil's puzzle and solution

3 (a) $w = 11$ (b) $d = 9$ (c) $k = 5$

4 The pupil's picture for
$10 + t = t + t + t + 2$, with $t = 4$

5 (a) $y = 4$ (b) $h = 13$ (c) $f = 5$
(d) $w = 12$

6 The pupil's picture for
$12 + 2q = q + 21$, with $q = 9$

Practice sheets

Sheet P70 (section B)

1 (a) 4   (b) 1

    (c) 6

2 9

3 4

4 3

5 4

6 2

7 6

Sheet P71 (section C)

1 (a) 3   (b) 0.5   (c) 4   (d) 50

    (e) 12   (f) 15   (g) 37

Sheet P72 (section D)

Pupils should write down their working and check.

1 (a) 3   (b) 0.5   (c) 12.5

2 The pupil's picture for
$y + y + 8 + y + y + y = 32 + y + y$
$y = 8$

3 (a) $s = 10$ (b) $w = 3$ (c) $m = 11$

    (d) $t = 9$ (e) $p = 49$ (f) $x = 33$

4 (a) $r = 8$ (b) $s = 6$ (c) $t = 3$

    (d) $u = 3$ (e) $v = 4$ (f) $w = 3$

Sheet P73 (section E) ▽□●

1 (a) $t = 6.5$ (b) $w = 6$ (c) $m = 12$

    (d) $t = 16$ (e) $p = 11$ (f) $x = 10$

2 (a) $t = 100$ (b) $w = 0.6$ (c) $m = 0.12$

    (d) $t = 160$ (e) $p = 110$ (f) $x = 1000$

Review 2 (p 91)

> **Essential**
> Angle measurer

1 (a) 16 (b) 8

2 The pupil's drawing for 2 purple tulips and 10 yellow tulips.

3

Number of purple tulips	1	2	3	4	5	6
Number of yellow tulips	8	10	12	14	16	18

4 (a) 26 (b) 12

5 (a) The number of yellows goes up in 2s.
  (b) The pupil's explanation

6 (a) 46
  (b) The pupil's explanation of how they worked it out

7 $(n \times 2) + 6$

8 (a) a and d are right angles, b is obtuse and c is acute.
  (b) $a = 90°$, $b = 145°$, $c = 35°$, $d = 90°$

9 (a) 4 (b) 14

10 (a) The pupil's solution and check with $n = 4$
  (b) The pupil's solution and check with answer 14

11 (a) The pupil's drawing
  (b) 85° (c) 540°

12 $a = 145°$ $b = 15°$ $c = 154°$
  $d = 26°$ $e = 122°$ $f = 132°$
  $g = 145°$ $h = 145°$ $i = 35°$
  $j = 251°$

Health club

This practical introduction to some of the ideas of data handling is based on a set of data cards. The cards can be sorted, ordered, arranged to make bar charts, and so on.

A lot of ideas are introduced in this unit. They are all taken up again in more detail later in the course.

p 93	**A** On record	▼■●	Using information on data cards
p 94	**B** On display	▼■●	Median (odd number of values)
			Dot plots, grouped bar charts
p 96	**C** Males and females	▼■●	Median (even number of values)
			Comparing two sets of data
p 97	**D** Healthy or not?	▽□●	Using a formula for body mass index

> **Essential**
> Sets of cards (one per pair of pupils) from sheets 80 and 81
> Sheets 82–87 (unless pupils draw their own axes for graphs)

A **On record** (p 93)

This is to familiarise pupils with the data cards.

> A set of cards per pair of pupils (from sheets 80 and 81)

◊ You could start by asking pupils why a health club would keep the kind of information on the cards (which would of course have to be updated periodically). Explain that rest pulse (in beats per minute) is measured when the person is sitting still. It is a rough measure of fitness (the lower the better).

◊ Ask some questions to help familiarise pupils with the information on the cards, such as 'Who weighs 60 kg?' or 'Who is the tallest female?'

◊ The first few follow-up questions are easy, but later ones are more difficult. It is not necessary for each pupil to do them all.

This introduces grouped frequency bar charts, median and dot plots.

> The cards, sheets 82, 83 (unless pupils draw their own axes for graphs)

◊ Ask pupils, in pairs, to sort the cards by age group. The cards can be placed to form a 'bar chart'. (You can do this on the board with Blu-tack.)

> 'I used a demo set of plastic coated cards. Pupils enjoyed putting them into grouped bar charts or on dot plots. Each child came and put a card on the board.'

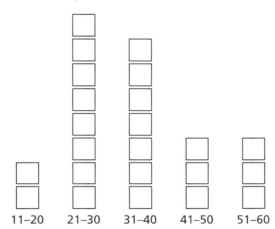

Pupils then draw the bar chart (on sheet 82, if used).

Discuss the distribution of age groups and the possible reasons for this.

Median

◊ Ask pupils to look at the heights on the cards, to put the cards in order of height and to find the middle height. (Some pupils may think of this as halfway between the heights of the shortest and tallest people. You can say that this is one interpretation of 'middle height' but not what you meant.)

Emphasise that the median is a height (168 cm), not a person. (It helps to use 'median' always as an adjective before 'height', 'weight', etc.) Emphasise also that there are equal numbers of people shorter than and taller than the median height.

Dot plot

◊ The cards themselves can be used to illustrate a dot plot of the heights. Place them on a large number line using Blu-tack:

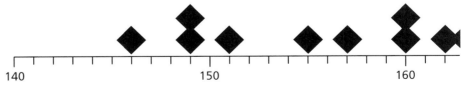

Pupils draw the dot plot on sheet 82.

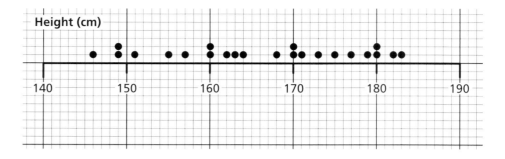

They could now discuss what the dot plot shows about the heights: they are fairly evenly spread but with more at the taller end.

Ask what sort of information can quickly be found from looking at the dot plot rather than the cards (for example, that the tallest person is 183 cm).

Grouped bar chart

◊ Compare the dot plot for heights with the bar chart for ages. The dot plot gives more detail but the bar chart shows the overall shape of the distribution better.

Discuss how to draw a bar chart for heights by grouping them.

Pupils draw a grouped bar chart on sheet 82. They can work directly from the cards or use the dot plot.

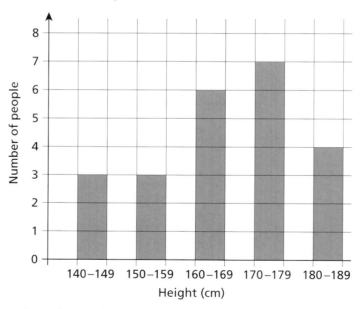

The bar chart shows clearly that most people at the club are between 160 and 179 cm tall.

Some pupils could look at the effect of choosing different class intervals.

B5 This question could be omitted by pupils who found B1 to B4 easy.

B8 This raises the issue of the median of an even number of values.

This is about comparing two sets of data.

> The cards, sheets 84 to 87

◊ Discuss the idea of comparing different groups of people. For example, the ages of males and females could be compared by drawing a bar chart for each (sheet 84).

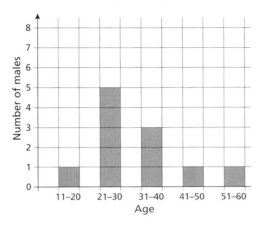
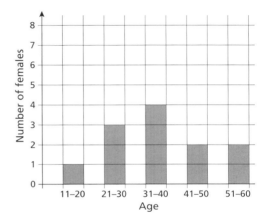

Comparing heights

◊ Discuss how the heights of the males and females could be compared. The median heights could be found. There are 12 females, so you will have to explain how to find the median in this case.

Finding the value halfway between two values is not as easy as it appears and the method of adding the two values and dividing by 2 can be baffling. It is more natural to halve the difference and add it to the smaller value. (A picture can help here.)

◊ Pupils can draw dot plots and grouped bar charts for the males and females on sheet 85.

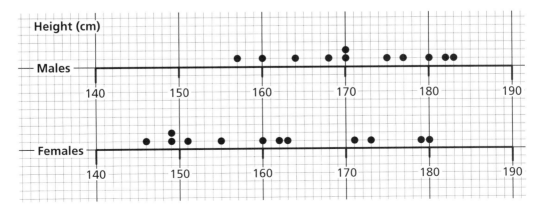

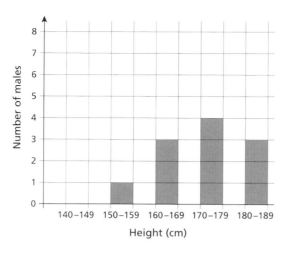
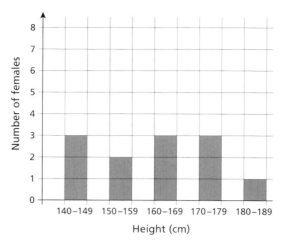

Ask what the diagrams show about the heights of males and females.

D Healthy or not? (p 97)

This uses the formula for *body mass index*.

> The cards

◊ Emphasise that the formula is valid only for adults. You may need to be sensitive to pupils' concerns about weight.

Body mass index is a crude measure and has been criticised by some medical authorities as misleading since it takes no account of bone, muscle, etc. as a proportion of mass.

Further work The cards can also be used to show a scatter diagram. For example, you could draw axes on the board for height and weight and Blu-tack each card at its position in the scatter diagram.

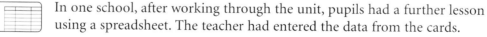

In one school, after working through the unit, pupils had a further lesson using a spreadsheet. The teacher had entered the data from the cards.

Pupils were shown how to sort the data, for example in ascending order of rest pulse. They also used the median command.

They were shown how to filter the data to select out all the females and find median values for this subgroup. They then tackled questions C1 and C3(a) using the spreadsheet.

A On record (p 93)

A1 81 kg

A2 74 bpm

A3 160 cm

A4 Any two from: J Abram, S Anderson,
U Borland, H Davis, J Dellano, G Giles,
M Lakhani, G Peters, H Tear

A5 K Berris – weight: 39 kg

A6 8 people

A7 J Abram

A8 A Saunders, W Stanwell

A9 K Berris, W Evans, E Pransch,
K Quaraishi, L Rogers, D Shah

A10 13 people

A11 H Tear

A12 5 males

A13 2 people

A14 M Lakhani, G Peters

B On display (p 94)

B1 64 kg

B2

B3

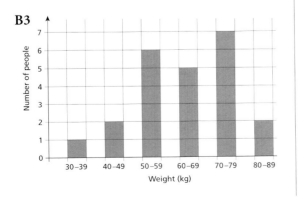

B4 The pupil's observations: for example,
most people at the club weigh between
50 kg and 80 kg and not many people
weigh over 80 kg.

B5 (a) 69 bpm

(b)

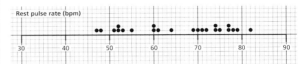

(c)

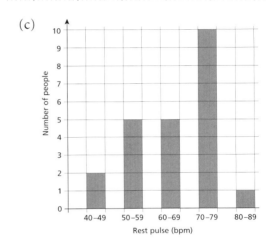

(d) The pupil's observations: for
example, almost half of the people at
the club have a rest pulse rate
between 70 and 79 bpm.

B6 (a) The pupil's explanations

(b) 32 years and 39 years are possible
median ages.

B7 22 years and 25 years are possible median
ages.

B8 (a) 167 cm

(b) 62.5 kg or $62\frac{1}{2}$ kg

C **Males and females** (p 96)

C1 (a) and (b)

Weights of males in order, lightest first:
46 kg, 60 kg, 63 kg, 64 kg, 70 kg, 70 kg,
72 kg, 74 kg, 76 kg 81 kg, 88 kg

Weights of females in order, lightest first:
39 kg, 46 kg, 50 kg, 53 kg, 55 kg, 56 kg,
58 kg, 59 kg, 66 kg, 66 kg, 70 kg, 79 kg

(c) 70 kg

(d) 57 kg

(e) The median weight for the males is
13 kg higher than the median weight
for the females suggesting that the
males at the health club tend to be
heavier than the females.

C2 (a)

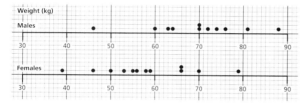

(b)

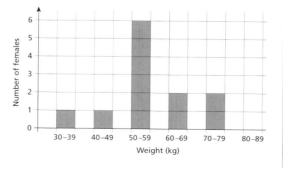

(c) The dot plots and bar charts support
the claim that the men at the club
tend to be heavier than the women.
The dots for the males are further
over to the right. On the bar chart
for the males, the bars tend to be
higher over to the right. The charts
show that all the males except one
weigh 60 kg or over but 8 out of the
12 females weigh below 60 kg.

C3 (a) The median rest pulse for the males
is 69 bpm.

The median rest pulse for the
females is 65 bpm.

(b)

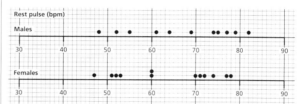

(c)

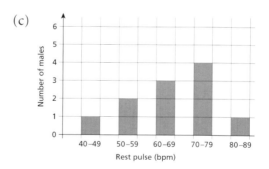

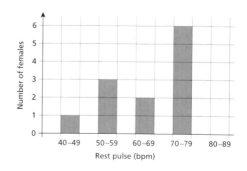

(d) The medians do not show a marked difference between the rest pulses for the males and females at the club. The dot plots show a fairly even spread for the males but a less even spread for the females where the dots tend to be clustered round two points. The bar chart for the females shows a 'dip' in the middle not shared by the graph for the males.

D **Healthy or not?** (p 97)

D1 Members of the club in the 11–20 age group are excluded as they may not be adults.

	Over a healthy weight	Under a healthy weight
Males	J Abram, S Anderson, L Unwin	
Females	W Evans, E Pransch, K Quaraishi	J Dellano, W Stanwell

Triangles

As this is a long unit, you may wish to break it into two parts, for example after section D.

Essential	Optional
Plain paper	Compasses for board or OHP
Compasses	Triangular dotty paper
Scissors	
Tracing paper	
An envelope to keep cut or traced triangles	
Thin card (for nets)	
Glue	
Angle measurer	
Sheet 88	
Practice sheets P74 to P77	

A Drawing a triangle accurately (p 98)

> Plain paper, compasses, tracing paper or scissors

◊ A good way to begin is by sketching an 8, 5, 10 triangle on the board and challenging the class to draw it accurately with pencil and ruler only. They may get an accurate result, but probably only after some trial and error. This should help them see the advantage of using compasses.

◊ If there are discrepancies when pupils compare their triangles with their neighbours', the problems should be sorted out and the triangles drawn again. They will be needed for later questions.

Investigation

For three lengths to make a triangle, the longest must be less than the sum of the other two.

B Equilateral triangles (p 100)

> The triangles made in section A, compasses, scissors, thin card, glue, (possibly) triangular dotty paper

◊ Pupils who, because of poor manipulative skills, are likely to be discouraged rather than helped by doing so much work with compasses could work on triangular dotty paper.

◊ Pupils could design and make other polyhedra with equilateral triangles as faces. Here is an easy-to-make net for a regular icosahedron.

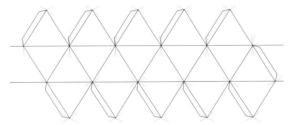

C Isosceles triangles (p 102)

> The triangles made in section A, sheet 88 (preferably on thin card), compasses, scissors, glue

◊ The fact that an equilateral triangle is a special case of an isosceles triangle may come up in the answers to questions and in discussion. There is no need to make a big thing of it at this stage.

C5 You may need to give help on naming triangles by the letters of their vertices.

D Scalene triangles (p 104)

> The triangles made in section A

◊ Ensure pupils realise that a right-angled triangle can be scalene.

E **Using angles** (p 105) ▼■●

This section includes drawing a triangle given one side and two angles, and given two sides and the angle between them.

> Angle measurer

F **Angles of a triangle** (p 108) ▽■●

> Angle measurer, scissors

◊ Ask everybody to draw a triangle, measure its angles and add them together. Enough of the results should be close to 180° to blame the discrepancy on inaccurate drawing and measurement!

◊ For the torn-off angles demonstration you may need to remind pupils about angles on a line.

Of course, neither of these approaches amounts to a proof. The proof is taken up in later material.

G **Using angles in isosceles triangles** (p 110) ▽■●

◊ The 'Can you make it?' box can be discussed with pupils when you draw together the work in the unit.

B **Equilateral triangles** (p 100)

B1 (a) Triangles C and F

(b) They are all 60°.

B2 Yes, you can fold it so two sides and two angles match up with one another. The fold line is the mirror line.

B3 6 vertices

B4 12 edges

B5 All the angles and sides of the triangular 'faces' are the same.

C **Isosceles triangles** (p 102)

C1 Triangles B, C, E and F (C and F are also equilateral).

C2 (a) It is 90° (a right angle).

(b) They are equal.

(c) It has reflection symmetry, with the fold line as a line of symmetry.

C3 The pupil's model

C4 Between 5 and 7 cm is a reasonable result (set squares are useful measuring aids). The important thing is that pupils don't just measure the sloping edge length or even the distance from the midpoint of the side of the square to the top of the vertex (a result of about 7.5 cm would indicate this).

C5 These triangles are isosceles. Finding just some of them is a fair achievement.

ACO, ECN, ACN, NCO, OCE, BOC, DNC, ANO, EON, FON, GNO, KGO, MFN, NIO, NLO, BCF, CDG, CFO, CGN, FIN, GIO

Diagonal cut puzzle

The triangles cut out are two pairs of isosceles triangles.

These are the two ways of fitting them together.

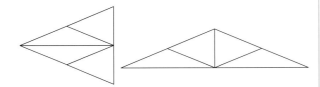

D Scalene triangles (p 104)

D1 Triangles A and D

D2 ABC, ABD, ACE, ADE

D3 (a) Isosceles
(b) Scalene
(c) Equilateral
(d) Isosceles
(e) Scalene
(f) Isosceles
(g) Equilateral

D4 No. Pupils can think about what would happen if they tried folding so that one side went on to another. They would be different lengths so they would not match.

E Using angles (p 105)

E1 The pupil's triangle

E2 The pupil's triangles

E3 It is not possible to complete the triangle because two of the lines are parallel.

E4 The pupil's triangle

E5 The pupil's triangles

E6 (a) The pupil's triangle
(b) BC = 7.5 cm, angle at B = 49°, angle at C = 61°

E7 (a) The pupil's triangle
(b) XZ = 7.4 cm, YZ = 12.4 cm, angle at Z = 20°

E8 (a) The pupil's triangle
(b) AC = 12.5 cm, angle at A = 61°, angle at C = 34°

E9 (a) The pupil's triangle
(b) QR = 12.5 cm, angle at Q = 32°, angle at R = 23°

E10 It is impossible to draw a triangle (because the circle centre B with radius 6 cm does not cross the line through A at 40° to AB).

E11 Two different triangles can be drawn (because the circle centre Q with radius 7 cm crosses the line through P at two points).

F Angles of a triangle (p 108)

F1 (a) 50° (b) 60° (c) 55°
(d) 80° (e) 20°

F2 (a) 75° (b) 121° (c) 85°
(d) 66° (e) 91°

F3 Each angle is 60°, because 180 ÷ 3 = 60.

F4 (a) 60° (b) 25° (c) 17°
(d) 62° (e) 33°

F5 $a = 60°$ $b = 120°$ $c = 70°$
$d = 110°$ $e = 80°$

F6 $a = 70°$ $b = 65°$ $c = 128°$
$d = 80°$ $e = 100°$ $f = 148°$

F7 To draw the triangle, the third angle (40°) must be worked out first.

Ⓖ Using angles in isosceles triangles
(p 110)

G1 They are equal.

G2 $a = 75°$ $b = 30°$ $c = 24°$
     $d = 132°$ $e = 54°$ $f = 72°$
     $g = 18°$ $h = 144°$ $i = 77°$
     $j = 26°$

G3 $a = 70°$ $b = 70°$ $c = 40°$
     $d = 40°$ $e = 66°$ $f = 66°$
     $g = 30°$ $h = 30°$ $i = 48°$
     $j = 48°$

G4 (a) Either 40° and 100° or 70° and 70°
     (b) Either 72° and 36° or 54° and 54°
     (c) 25° and 25°

G5 The triangles will not fit together because
     50° does not go into 360° exactly.

G6 9 sides

G7 30°

G8 144°

What progress have you made? (p 112)

1 The pupil's triangle

2 (a) ABD
     (b) EBD, ABD
     (c) ABE, ADE, BEC, DEC, ABC, ADC
     (d) ABC, EBC, ADC, EDC

3 The pupil's triangles

4 $a = 112°$ $b = 57°$

5 $a = 67°$ $b = 46°$ $c = 38°$

Practice sheets

Sheet P74 (sections A and D)   ▼■○

1 The pupil's triangle
    The angles should be 87°, 51°, 42°.

2 The pupil's triangle
    The angles should be 90°, 53°, 37°.

3 (a) C, D
     (b) A, E
     (c) B, F, G
     (d) B, F

Sheet P75 (section E)   ▼■●

1 The pupil's triangle
    The third angle should be 80°.

2 The pupil's triangle
    The third angle should be 105°.

3 The pupil's triangle
    The third side should be 4.5 cm.

4 The pupil's triangle
    The third side should be 11.5 cm.

Sheet P76 (section F)   ▽■●

 $a = 60°$ $b = 40°$ $c = 25°$
 $d = 58°$ $e = 139°$ $f = 65°$
 $g = 115°$ $h = 49°$ $i = 131°$
 $j = 38°$ $k = 67°$ $l = 43°$

Sheet P77 (section G)   ▽■●

1 $a = 74°$ $b = 74°$ $c = 55°$
     $d = 70°$ $e = 22°$ $f = 22°$
     $g = 40°$

2 Yes, its angles are 90°, 45°, 45°.

3 Either 50° and 80° or 65° and 65°

4 (a) 45°   (b) $67\frac{1}{2}°$

5 $a = 40°$ $b = 100°$ $c = 80°$
     $d = 110°$ $e = 10°$ $f = 80°$
     $g = 70°$ $h = 137°$ $i = 111°$
     $j = 138°$ $k = 75°$

Work to rule

The emphasis in this unit is on finding rules by analysing tile designs. Some pupils may find the concrete experience of making the patterns with tiles or multilink helpful.

No use is made of the convention that multiplication and division have priority over addition and subtraction: brackets are used throughout.

T

p 113	**A** Mobiles	▼■●	Finding and using a rule to calculate the number of white tiles given the number of red tiles
p 116	**B** Explaining	▽■●	Considering explanations of how rules were found
p 118	**C** Towers and L-shapes	▼□○	Finding and using a rule that uses one operation and describing it in words
p 120	**D** Bridges	▼■○	Finding and using a rule that uses one operation and describing it in words
p 122	**E** Surrounds	▼■○	Finding and using a rule that uses two operations and describing it in words
p 123	**F** More designs	▼□○	Investigating designs with linear rules

T

p 124	**G** Shorthand	▽■●	Understanding algebraic shorthand: $w = 2r + 1$
p 125	**H** Using shorthand	▽■●	Finding and using rules such as $w = 4r + 3$
p 127	**I** T-shapes and more	▽□●	Finding and using rules such as $w = \frac{r}{2} + 3$
p 129	**J** Snails and hats	▽■●	Considering designs that give quadratic rules

---

Optional

Tiles or multilink in two colours
Squared paper
Sheet 89

Practice sheets P78 to P80

In this section pupils should become aware that there are two types of rule for these patterns:

- by tabulating results in order we can see how the sequence of white tiles continues – it goes up in 2s
- by looking at the structure of the pieces we can see that the number of whites is equal to double the number of reds plus 3

Pupils should begin to appreciate the advantages of the latter rule.

> Optional: Tiles or multilink in two colours, squared paper

◊ You can start by discussing the design of the pieces in the mobile on page 113, perhaps using tiles or multilink.

Tabulate results in order up to, say, 8 red tiles, and look at the pattern in the table. Many pupils will spot that the number of white tiles goes up in 2s, but they should think about why it will continue in 2s.

Ask pupils to imagine the piece with, say, 100 red tiles and how we could calculate the number of white tiles. Pupils should appreciate it would take a long time to continue to add on 2s.

A discussion of the structure of the designs should lead to

- the piece with 100 red tiles has $(100 \times 2) + 3$ whites

and to the general rule

- to find the number of whites, multiply the number of reds by 2 and add 3

All but less able pupils should be able to grasp the shorthand version

number of whites = (number of reds $\times$ 2) + 3

Use the discussion to bring out how diagrams can be useful in making the structure of the designs clear, for example:

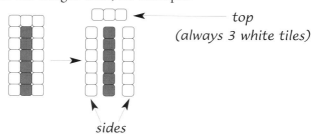

top
(always 3 white tiles)

sides

◊ When asked to explain how to find the number of white tiles for a given number of reds, some pupils may use repeated addition. Encourage all pupils to think about a rule to calculate directly the number of whites in some way. Some pupils will be able to see from the table that because the number of whites 'goes up in 2s' we must multiply the number of reds by 2 when seeking a direct rule.

'The pupils were very comfortable with the work. I was very pleased with their explanations.'

A7 Pupils who give 102 white tiles as their answer may have used the rule '+ 2' to continue the sequence. Others may try to add on 2s. Encourage them to visualise the piece with 100 reds to enable them to see that the result can be found by multiplying by 2 and adding 1.

A10 Pupils who choose 'number of white tiles = number of red tiles + 2' may be confusing the rule to continue the sequence with the rule to calculate the number of whites given the number of reds.

B **Explaining** (p 116)

◊ This section may be used as the basis for small group or whole-class discussion. The explanations are not intended to serve as exemplars: encourage pupils to explain in their own way.

C **Towers and L-shapes** (p 118) ▼□○

C7 Lower attainers often find it difficult to give written explanations, but they may be able to explain orally why you multiply the number of reds by 2 to get the number of whites.

D **Bridges** (p 120)

'Went well – but they didn't think to draw diagrams themselves for the table in D5.'

D7 Pupils may have counted on in 1s to answer question D6. Ask them to visualise the piece with 20 reds to help them see that the result can be found by adding 4 to the number of red tiles.

D9 Encourage pupils to draw diagrams to illustrate their explanations. Some pupils will find this difficult and should be encouraged to offer their own explanations, however tentative. An oral explanation would be perfectly acceptable at this stage.

E **Surrounds** (p 122) ▼■○

E7 Pupils may produce equivalent rules, for example:
number of white tiles = ((number of red tiles + 2) × 2) + 2
number of white tiles = number of red tiles + number of red tiles + 6

This is an opportunity to show equivalence with the rule
number of white tiles = (number of red tiles × 2) + 6

F **More designs** (p 123)

These questions are intended to provide pupils with an opportunity to structure their own work to find rules. Encourage them to use strategies such as counting tiles in the examples given, drawing more diagrams and tabulating results to help them analyse the diagrams to find a rule.

Pupils could select one or more of these designs to investigate. They could work in groups and present their ideas to the whole class.

G **Shorthand** (p 124)

H **Using shorthand** (p 125)

◊ Some pupils may feel more confident with rules in the form $w = (r \times 2) + 8$ rather than $w = 2r + 8$. Both forms are acceptable answers.

◊ Encourage all pupils, especially the more able, to give written explanations, although oral explanations are acceptable at this stage.

H7 This is intended to provide pupils with an opportunity to structure their own work to find rules. Encourage them to use strategies such as counting tiles in the examples given, drawing more diagrams and tabulating results to help them analyse the diagrams to find a rule.

Pupils could work in groups and present their ideas on one or more of these designs to the whole class.

I **T-shapes and more** (p 127)

I1 Pupils can select one or more of these designs to investigate. They could work in groups and present their ideas to the whole class.

J **Snails and hats** (p 129)

> Optional: Sheet 89

◊ Some pupils may need help in seeing how the snails are made.

J8 (b) Pupils are likely to find this difficult.

J9 (b) Some pupils may choose to add 13 to the result for 6 red tiles; others may calculate $(7 \times 7) + 1$. This gives an opportunity to compare these two methods.

*J18 This question is on sheet 89 and will extend even the most able.

A Mobiles (p 113)

A1 5 white tiles

A2 (a) The pupil's piece with 3 reds

(b) 7 white tiles

A3 (a) The pupil's piece with 5 reds

(b) 11 white tiles

A4

Number of red tiles	1	2	3	4	5	6
Number of white tiles	**3**	**5**	**7**	9	**11**	**13**

A5 (a) As the number of red tiles goes up by 1, the number of white tiles goes up by 2 each time.

(b) The pupil's explanations: for example, an increase of 1 red tile means an extra 2 white tiles, one on each side.

A6 (a) 17 white tiles (b) 21 white tiles

A7 201 white tiles

A8 The pupil's method: for example, to find the number of white tiles, multiply the number of red tiles by 2 and add 1; or a method that involves repeated addition of 2.

A9 301 white tiles

A10 number of white tiles = (number of red tiles × 2) + 1

A11 (a) 49 white tiles (b) 145 white tiles

A12 9 red tiles

A13 40 red tiles

A14 The pupil's explanation: for example, the number of white tiles is always odd.

B Explaining (p 116)

B1 (a) The pupil's response

(b) The pupil's response

B2 The pupil's response

B3 The pupil's response and explanation

C Towers and L-shapes (p 118)

C1 10 white tiles

C2 (a) The pupil's tower with 6 reds

(b) 12 white tiles

C3

Number of red tiles	1	2	3	4	5	6
Number of white tiles	**2**	4	**6**	**8**	**10**	**12**

C4 (a) 16 white tiles

(b) The pupil's tower with 8 reds

C5 30 white tiles

C6 80 white tiles

C7 The pupil's method: for example, multiply the number of red tiles by 2; or a method that involves repeated addition of 2.

C8 (a) 200 white tiles

(b) 116 white tiles

C9 (a) The pupil's L-shape with 5 reds

(b) 7 white tiles

C10

Number of red tiles	1	2	3	4	5	6
Number of white tiles	**3**	4	**5**	**6**	**7**	**8**

C11 (a) 11 white tiles

(b) The pupil's L-shape with 9 reds

C12 28 white tiles

C13 102 white tiles

C14 The pupil's method: for example, add 2 to the number of red tiles.

D Bridges (p 120)

D1 10 white tiles

D2 (a) The pupil's bridge with 2 reds

(b) 6 white tiles

D3 (a) The pupil's bridge with 4 reds

(b) 8 white tiles

D4 3 red tiles

D5

Number of red tiles	1	2	3	4	5	6
Number of white tiles	5	6	7	8	9	10

D6 (a) 14 white tiles

(b) The pupil's bridge with 10 reds

D7 24 white tiles

D8 54 white tiles

D9 1004 white tiles

D10 The pupil's method: for example, add 4 to the number of red tiles.

D11 (a) 29 white tiles (b) 40 white tiles

D12 (a) number of white tiles = number of red tiles + 4 or

number of white tiles = 4 + number of red tiles

(b) The pupil's explanation

D13 (a) 87 white tiles (b) 100 white tiles

D14 (a) 48 red tiles

(b) The pupil's method: for example, 52 − 4 = 48

E Surrounds (p 122)

E1 (a) The pupil's surround with 6 reds

(b) 18 white tiles

E2

Number of red tiles	1	2	3	4	5	6
Number of white tiles	8	10	12	14	16	18

E3 (a) 26 white tiles

(b) The pupil's surround with 10 reds

E4 58 white tiles

E5 206 white tiles

E6 The pupil's explanation

E7 (a) Any rule equivalent to
number of white tiles =
(number of red tiles × 2) + 6

(b) The pupil's explanation

F More designs (p 123)

F1 (a) Set A: 102 white tiles

Set B: 202 white tiles

Set C: 401 white tiles

Set D: 204 white tiles

(b) The pupil's explanations of how to find the number of whites in each set.

G Shorthand (p 124)

G1 A and F, B and H, C and E, D and G

G2 Any letters can be used to stand for the *number of white tiles* and the *number of red tiles*. We have used w for the *number of white tiles* and r for the *number of red tiles*.

(a) $w = 4r + 3$ or equivalent

(b) $w = 5r$ or equivalent

(c) $w = 2r + 7$ or equivalent

(d) $w = r + 6$

(e) $w = 3r - 1$ or equivalent

H Using shorthand (p 125)

H1 (a) 23 white tiles

(b) The pupil's description of a check, possibly involving drawing the piece with 5 tiles

H2 (a) 51 white tiles

(b) 243 white tiles

H3 (a) Any rule equivalent to
number of white tiles =
(number of red tiles × 4) + 3

(b) $w = 4r + 3$ or equivalent

(c) The pupil's explanation of why the
rule works

H4 (a) 103 white tiles

(b) 163 white tiles

H5 (a) The pupil's design that fits
$w = 2r + 2$

(b) The pupil's explanation (with
diagrams)

H6 (a) $w = r + 1$

(b) The pupil's explanation of why the
rule works

H7 For each rule (or its equivalent) the pupil
should give an explanation of why it
works.

Set A: $w = r + 4$

Set B: $w = 3r + 1$

Set C: $w = 5r + 3$

T-shapes and more (p 127)

I1 For each rule (or its equivalent) the pupil
should give an explanation of why it
works.

Set A: (a) $w = \frac{r}{2} + 6$ (c) 56 white tiles

Set B: (a) $w = r - 2$ (c) 98 white tiles

Set C: (a) $w = \frac{r}{2} + 3$ (c) 53 white tiles

Set D: (a) $w = r + 4$ (c) 104 white tiles

Set E: (a) $w = 2r - 4$ (c) 196 white tiles

Set F: (a) $w = \frac{3r}{2} + 5$ (c) 155 white tiles

Snails and hats (p 129)

J1 10 white tiles

J2 (a) The pupil's snail with 4 red tiles

(b) 17 white tiles

J3 37 white tiles

J4 101 white tiles

J5 401 white tiles

J6 The pupil's explanation

J7

Number of red tiles	1	2	3	4	5	6
Number of white tiles	**2**	5	**10**	**17**	**26**	**37**

J8 (a) The pupil's description: for example,
the number of white tiles goes up
by 3, then 5, then 7, then 9, … –
these numbers go up by 2 each time.

(b) The pupil's explanation: for example,
each snail is made by adding an L-
shape of tiles to the previous one and
the L-shape gets bigger by 2 each
time.

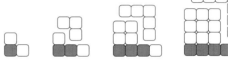

J9 (a) 50 white tiles

(b) The pupil's method: for example,
$37 + 13 = 50$, $(7 \times 7) + 1 = 50$

J10 12 red tiles

J11 4 white tiles

J12 (a) The pupil's hat with 6 reds

(b) 16 white tiles

J13 100 white tiles

J14 The pupil's explanation

J15 529 white tiles

J16 10 red tiles

J17 The pupil's explanation

Sheet 89

A (a) 102 white tiles

(b) 10 002 white tiles

(c) $w = g^2 + 2$

(d) 2502 white tiles

B Pupils may find equivalents to these answers.

Set A: $w = \left(\dfrac{g}{4}\right)^2 + 4$

Set B: $w = g^2 + 2g + 4$

Set C: $w = \left(\dfrac{g-1}{2}\right)^2$

(g has to be an odd number.)

Set D: $w = g(g-1)$ or $w = g^2 - g$

Set E: $w = 3g + 4$

What progress have you made? (p 130)

1 8 white tiles

2 (a) The pupil's drop with 5 red tiles

   (b) 9 white tiles

3

Number of red tiles	1	2	3	4	5
Number of white tiles	**5**	**6**	7	**8**	**9**

4 (a) 11 white tiles

   (b) 14 white tiles

5 104 white tiles

6 The pupil's explanation

7 17 white tiles

8 302 white tiles

9 The pupil's explanation

10 33 white tiles

11 (a) $w = r + 8$

   (b) $w = 4r + 1$ or $w = (r \times 4) + 1$

12 (a) $w = 2r + 4$ or equivalent

   (b) The pupil's explanation

   (c) 104 white tiles

13 Set X

   (a) $w = r - 1$ or equivalent

   (b) The pupil's explanation

   (c) 99 white tiles

Set Y

(a) $w = \dfrac{r}{2} - 2$ or equivalent

(b) The pupil's explanation

(c) 48 white tiles

Practice sheets

Sheet P78 (section A)   ▼■●

1 3 whole circles

2 7 whole circles

3 (a) The pupil's frieze

   (b) 9 whole circles

4

Number of tiles	1	2	3	4	5	6
Number of circles	**1**	**3**	5	**7**	**9**	**11**

5 (a) The number of circles goes up by 2 each time a tile is added.

   (b) The pupil's explanation: for example, each time a tile is added one half circle on the join is completed and one circle in the centre is added.

6 (a) 15 circles   (b) 19 circles

7 199 circles

8 The pupil's explanation: for example, double the number of tiles and subtract 1.

9 299 circles

10 number of circles = (number of tiles $\times$ 2) $-$ 1

11 (a) 53 circles   (b) 161 circles

12 13 tiles

13 45 tiles

14 The pupil's explanation: for example, the number of circles is always an odd number

Sheet P79 (sections D, F) ▼■○

1 16 whole circles

2 (a) The pupil's frieze   (b) 4 circles

3 (a) The pupil's frieze   (b) 13 circles

4 19 circles

5

Number of tiles	1	2	3	4	5	6
Number of circles	**1**	**4**	7	**10**	**13**	**16**

6 (a) 28 circles   (b) The pupil's frieze

7 58 circles

8 148 circles

9 The pupil's explanation, such as 'It's the number of tiles plus twice one less than the number of tiles.'

10 (a) 73 circles   (b) 124 circles

11 (a) number of whole circles = number of tiles + 2 × (number of tiles − 1)

(b) The pupil's explanation: for example, looking at the friezes, there is a middle row of circles (one on each line) plus a top and a bottom row (one circle on each join line).

12 (a) 259 circles   (b) 283 circles

13 (a) 20 tiles

(b) The pupil's method: for example, '20 is a bit more than a third of 58, so I tried 20 and it worked.'

14 Tile A

(a) 199 circles

(b) There will be one row of 100 circles in the centres of the tiles, plus another row of 99 circles along the top.

(c) number of whole circles = number of tiles + (number of tiles − 1) or equivalent

Tile B

(a) 299 circles

(b) There are two whole circles in each tile (200) plus another 99 made of half circles.

(c) number of whole circles = (number of tiles × 3) − 1 or equivalent

Sheet P80 (section H) ▽■●

1 (a) 8 circles   (b) Draw a sketch.

2 (a) 14 circles   (b) 44 circles

3 (a) number of whole circles = number of tiles + (($\frac{1}{2}$ × number of tiles) − 1)

(b) $c = t + \frac{t}{2} - 1$ or $c = \frac{3t}{2} - 1$ or equivalent

(c) There is one circle in the centre of each tile plus a middle row of circles on the joins.

4 (a) 35   (b) 74

5 Set A
$c = 3t - 3$
There are 3 rows of circles; the number of circles in each row is one less than the number of tiles.

Set B
$c = \frac{5t}{2} - 2$
There are 2 rows of circles in which the number of circles is one less than the number of tiles; there is also a middle row on the joins in which the number of circles is half the number of tiles.

Area and perimeter

Most of the work involves rectangles. There is introductory work on right-angled triangles.

Essential
Centimetre squared paper

Practice sheets P81 to P85

A Exploring perimeters (p 133) ▼■●

◊ These investigations are arranged in increasing difficulty. Pupils can go as far with them as they can manage, but all should become familiar with the word *perimeter* and should come to see that a particular number of squares (producing shapes of a fixed area) can give rise to different perimeters. The perimeters are all even numbers of centimetres (investigation 3) because before the squares are put together their total perimeter is an even number of centimetres (actually a multiple of 4) and each time two edges are put together the total perimeter falls by 2 centimetres, remaining even.

The chart on the next page shows the possible perimeters for investigations 2, 4 and 5.

The maximum perimeter for a given number of squares n is $2n + 2$ or $2(n + 1)$ and arises when all the squares are in a straight line or form a shape 'one square wide' with bends in it (the least 'compact' arrangement). Some pupils should be able to explain, in their own words, why such a relationship applies. This linear relationship is shown by the upper right-hand edge of the block of ticks on the chart.

Perimeter

Number of squares	4	6	8	10	12	14	16	18	20	22	24	26	28	30	32	34	36
1	✓																
2		✓															
3			✓														
4			✓	✓													
5				✓	✓												
6				✓	✓	✓											
7					✓	✓	✓										
8					✓	✓	✓	✓									
9					✓	✓	✓	✓	✓								
10						✓	✓	✓	✓	✓							
11						✓	✓	✓	✓	✓	✓						
12						✓	✓	✓	✓	✓	✓	✓					
13							✓	✓	✓	✓	✓	✓	✓				
14							✓	✓	✓	✓	✓	✓	✓	✓			
15							✓	✓	✓	✓	✓	✓	✓	✓	✓		
16							✓	✓	✓	✓	✓	✓	✓	✓	✓	✓	
17								✓	✓	✓	✓	✓	✓	✓	✓	✓	✓

The minimum perimeter is shown by the left-hand edge of the block of ticks. This relationship is not linear but pupils may be able to give their own explanations of why the steps shown with arrows in the chart occur.

B **Square centimetres** (p 134)

C **The area of a rectangle** (p 135)

D **Measuring to find areas** (p 136)

◊ Some pupils may need support with questions D3 and D4, where the process is inverted.

E **It pays to advertise** (p 138)

◊ This is straightforward practice which those on the ▼ path can use while you give your attention to those working on section G.

F Square metres (p 140)

◊ Pupils often calculate answers in square metres with no idea how big a square metre is. At this stage, it is worthwhile making a metre square from newspaper and referring to it in working out or estimating the area of, say, the classroom wall.

It is also worth doing some work on estimating, say, the amount (and cost) of grass seed or turf needed to make an actual lawn (or sports field!). In 1999, a 1 kg box of grass seed sufficient for 20 square metres of grass cost about £6 to £8, and turf cost about £1.75 per square metre.

◊ From F4 onwards, pupils need to be systematic about working out missing lengths of sides. A sketch diagram is always a good idea and should be encouraged.

G Bringing in triangles (p 144)

H Using decimals (p 145)

◊ This section assumes confidence with the idea that a half is 0.5 and a quarter is 0.25. It aims to help pupils see that the result of a decimal multiplication corresponds to the total of whole, half and quarter squares in an area diagram. You may need to check that pupils are following the ideas, through teacher-led class discussion.

I Further decimals (p 147)

'Question I20 promoted good discussion in splitting up the shapes.'

◊ This section aims not only to promote competence in working out areas from decimal lengths but also to use area to give pupils an insight into how multiplication works (for example, in I6, to show why the common 'result' $0.4 \times 0.2 = 0.8$ is wrong). So pupils must think their own way through questions I1 to I10, using the calculator only to check. You may need to provide some additional examples to fix the ideas, and you will almost certainly need to go through 'the Ritz Hotel's big doormat' (top of page 150) on the board and possibly one or two more like it. From question I11 onwards pupils can use their calculators to obtain results. This work complements the price and quantity approach to decimal multiplication used in Unit 7 'Calculating with decimals' in *Book N*.

True or false? (p 153)

Discussion of these questions may help to clear up common misunderstandings.

Safely grazing (p 154)

This investigation may be done at any suitable time during the unit.

For pupils who have become confident with decimals, 'Safely grazing' can be adapted to start from a total fence length that gives rise to non-integer sides for the pen.

B Square centimetres (p 134)

B1 A 7 cm^2, B 8 cm^2, C 5 cm^2

B2 (a) A has the largest area (15 cm^2).

(b) B has the smallest area (13 cm^2).

(c) C and E have the same area (14 cm^2).

B3 The pupil's shapes

C The area of a rectangle (p 135)

C1 12 cm^2; the pupil's drawings of 1 cm by 12 cm, 2 cm by 6 cm rectangles

C2 The pupil's three rectangles such as
1 cm by 40 cm
2 cm by 20 cm
4 cm by 10 cm
5 cm by 8 cm

C3 1 cm by 16 cm
2 cm by 8 cm
4 cm by 4 cm (You can point out that a square is a special case of a rectangle.)

C4 (a) 18 cm^2 (b) 28 cm^2 (c) 25 cm^2
(d) 12 cm^2 (e) 30 cm^2

D Measuring to find areas (p 136)

D1 (a) 15 cm^2 (b) 24 cm^2 (c) 20 cm^2
(d) 8 cm^2 (e) 16 cm^2 (f) 12 cm^2
(g) 18 cm^2 (h) 21 cm^2

D2 (a) The pupil's answer (many people think more than half is shaded).

(b) The area of the whole rectangle is 42 cm^2; the area of the shaded part is 20 cm^2. So less than half is shaded.

D3 (a) 7 cm (b) 7 cm (c) 4 cm
(d) 8 cm

D4 (a) 20 cm (b) 15 cm (c) 60 cm
(d) 6 cm

D5 27 cm^2

E It pays to advertise (p 138)

E1 (a) £35 (b) £48 (c) £30
(d) £25 (e) £30 (f) £36

E2 (a) £28 (b) £64 (c) £84
(d) £20 (e) £80 (f) £72

F Square metres (p 140)

F1 50 m^2

F2 (a) 8 m^2 (b) 4 m^2 (c) 12 m^2

F3 The pupil's sketches leading to a total area of 12 m^2

F4 (a) 14 m^2 (b) 19 m^2 (c) 15 m^2
(d) 42 m^2 (e) 18 m^2

F5 (a) 16 m (b) 22 m (c) 18 m
(d) 28 m (e) 22 m

F6 A 240 m^2, B 144 m^2, C 280 m^2
D 240 m^2, E 400 m^2, F 24 m^2
G 480 m^2, H 112 m^2, I 72 m^2

In each case, the total area is 664 m^2.

F7 (a) 1275 m^2 (b) 2256 m^2
(c) 1975 m^2 (d) 1400 m^2

F8 A 352 m^2, B 48 m^2, C 63 m^2
So the area of the black shape is 241 m^2.

F9 (a) 582 m^2 (b) 480 m^2 (c) 772 m^2

Bringing in triangles (p 144)

G1 (a) A half (b) $3\,cm^2$

G2 (a) $10\,cm^2$ (b) $7\,cm^2$ (c) $1\,cm^2$
(d) $1.5\,cm^2$ (e) $7.5\,cm^2$

G3 (a) $7.5\,cm^2$ (b) $13\,cm^2$ (c) $5\,cm^2$
(d) $6\,cm^2$ (e) $7\,cm^2$ (f) $4\,cm^2$
(g) $2\,cm^2$

G4 (a) $6\,cm^2$ (b) $8\,cm^2$ (c) $6\,cm^2$

ℍ **Using decimals** (p 145)

H1 (a) $10.5\,cm^2$ (b) $9\,cm^2$ (c) $7\,cm^2$

H2 (a) $6\,cm^2$ (b) $12.5\,cm^2$ (c) $10\,cm^2$
(d) $4.5\,cm^2$ (e) $18\,cm^2$ (f) $22.5\,cm^2$

H3 (a) A quarter (b) 0.25 (c) $8.75\,cm^2$

H4 (a) $5.25\,cm^2$ (b) $11.25\,cm^2$
(c) $15.75\,cm^2$

H5 (a) $3.75\,cm^2$ (b) $13.75\,cm^2$
(c) $22.75\,cm^2$ (d) $0.75\,cm^2$
(e) $6.25\,cm^2$ (f) $8.25\,cm^2$

H6 (a) $4\,cm$ (b) $9\,cm$ (c) $18\,cm$
(d) $12\,cm$ (e) $36\,cm$

H7 (a) $11\,cm$ (b) $15\,cm$ (c) $13\,cm$
(d) $9\,cm$ (e) $17\,cm$ (f) $19\,cm$

H8 (a) $8\,cm$ (b) $16\,cm$ (c) $20\,cm$
(d) $4\,cm$ (e) $10\,cm$ (f) $14\,cm$

H9 (a) $3\,cm$ by $1.5\,cm$
(b) $2.5\,cm$ by $2\,cm$
(c) $4\,cm$ by $2.5\,cm$
(d) $5.5\,cm$ by $3.5\,cm$

𝕀 **Further decimals** (p 147)

I1 (a) $0.6\,m^2$ (b) $0.4\,m^2$ (c) $0.3\,m^2$

I2 (a) $0.05\,m^2$ (b) $0.09\,m^2$ (c) $0.06\,m^2$

I3 (a) $0.82\,m^2$ (b) $0.48\,m^2$ (c) $0.32\,m^2$

I4 (a) 24 (hundredths of a square metre)
(b) $0.24\,m^2$
(c) The pupil's check

I5 A: (a) 35 (b) $0.35\,m^2$
B: (a) 72 (b) $0.72\,m^2$
C: (a) 40 (b) $0.40\,m^2$ (or $0.4\,m^2$)

I6 (a) $0.18\,m^2$ (b) $0.42\,m^2$ (c) $0.10\,m^2$

I7 No, a doormat with these dimensions
would cover 8 hundredths of a square
metre, which is $0.08\,m^2$ as a decimal.
Emma's answer, 0.8, would give 8 **tenths** of
a square metre, nearly a full square metre.

I8 (a) $0.06\,m^2$ (b) $0.05\,m^2$ (c) $0.12\,m^2$
(d) $0.20\,m^2$ (e) $0.40\,m^2$ (f) $0.5\,m^2$

I9 (a) 160 (hundredths of a square metre)
(b) $1.60\,m^2$
(c) The pupil's check

I10 (a) $1.20\,m^2$ (b) $1.80\,m^2$

I11 The pupil's sketches and working, leading
to
A $6.82\,m^2$, B $5.40\,m^2$, C $8.68\,m^2$

I12 (a) $11.48\,cm^2$ (b) $8\,cm^2$ (c) $16.2\,cm^2$

I13 Answers may differ slightly because of
printing distortion.
(a) £32.68 ($7.6\,cm$ by $4.3\,cm$)
(b) £25.30 ($4.6\,cm$ by $5.5\,cm$)
(c) £52.06 ($13.7\,cm$ by $3.8\,cm$)

I14 (a) $23.57\,cm^2$ (b) $18.92\,cm^2$

I15 (a) $3.5\,m$ (b) $2.1\,m$ (c) $8.1\,m$

I16 $7.5\,cm$ and $4.8\,cm$

I17 (a) $9.88\,cm^2$ (b) $4.9\,cm^2$
(c) $8.405\,cm^2$

I18 (a) $9.86\,cm^2$ (b) $17.36\,cm^2$

I19 Measurements may differ slightly because of printing distortion. Lengths are in centimetres.

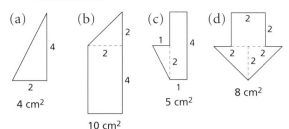

(a) 4 cm²

(b) 10 cm²

(c) 5 cm²

(d) 8 cm²

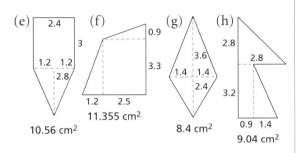

(e) 10.56 cm²

(f) 11.355 cm²

(g) 8.4 cm²

(h) 9.04 cm²

True or false? (p 153)

• Shapes do not have to have straight sides to have an area.

• Curved surfaces also have areas.

• Some map projections do keep area constant, but the common Mercator's projection does not – areas appear bigger than they really are as you go north or south from the Equator.

Safely grazing (p 154)

The biggest pen is 10 m by 10 m, with an area of 100 m².

What progress have you made? (p 154)

1 Area 24 cm², perimeter 22 cm

2 The pupil's sketch giving area 16 m²

3 (a) Area 3.75 cm², perimeter 8 cm
   (b) Area 12.25 cm², perimeter 14 cm

4 (a) 7.5 cm²   (b) 3 cm²
   (c) 2 cm²   (d) 3.5 cm²

5 (a) 8.4 cm²   (b) 1.98 cm²

Practice sheets

Sheet P81 (sections B, C, E)   ▼□○

1 A 7 cm², B 4 cm², C 7 cm², D 2 cm²

2 (a) 16 cm²   (b) 18 cm²   (c) 15 cm²
   (d) 4 cm²   (e) 10 cm²

3 The pupil's three rectangles with area 24 cm²

4 (a) 2 cm by 4 cm, 8 cm²
   (b) 3 cm by 7 cm, 21 cm²
   (c) 6 cm by 2 cm, 12 cm²
   (d) 4 cm by 7 cm, 28 cm²
   (e) 3 cm by 2 cm, 6 cm²
   (f) 5 cm by 4 cm, 20 cm²
   (g) 3 cm by 8 cm, 24 cm²

Sheet P82 (section F)   ▼■●

1 The pupil's sketches, giving areas
   (a) 17 m²,   (b) 14 m²,   (c) 14 m²,
   (d) 19 m²,   (e) 20 m²,   (f) 10 m²

Sheet P83 (section F)   ▽■●

1 The pupil's sketches, giving areas
   (a) 324 m²,   (b) 1400 m²,   (c) 706 m²,
   (d) 231 m²,   (e) 210 m²,   (f) 72 m²,
   (g) 252 m²

Sheet P84 (section G)   ▽■●

1 (a) 5 cm²   (b) 5.25 cm²
   (c) 20.25 cm²

2 (a) 6.25 cm²   (b) 16.5 cm²
   (c) 29.25 cm²   (d) 9 cm²
   (e) 11.25 cm²   (f) 11.25 cm²

3 (a) 7.5 cm   (b) 8.5 cm
   (c) 4.5 cm   (d) 5.5 cm

1 The pupil's working with answers
   (a) 0.35 m²   (b) 0.36 m²   (c) 0.4 m²

2 (a)

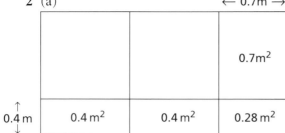

   (b) Total area = 3.78 m²

3 (a) 2.4 m², 6.2 m
   (b) 6 m², 9.8 m
   (c) 2.7 m², 6.6 m
   (d) 6.51 m², 10.4 m
   (e) 32.4 m², 23.4 m
   (f) 3.1 m², 13.4 m

4 (a) 4.5 cm by 4 cm
   (b) 2.5 cm by 7.5 cm
   (c) 1.6 cm by 3.4 cm
   (d) 0.5 cm by 8 cm

Review 3 (p 156)

1 A (0, 3), B (4, 6), C (10, 3), D (4, 0)

2 4

3 You also need compasses.
   The pupil's 6, 7, 10 cm triangle

4 (a) 20 cm   (b) 21 cm^2

5 The distance between the two points should be between 4.1 and 4.2 cm.

6 (a) The pupil's drawing; 5 white tiles
   (b)

Number of red tiles	1	2	3	4	5	6	7
Number of white tiles	**2**	**3**	4	**5**	**6**	**7**	**8**

   (c) (i)   11   (ii) 51
   (d), (e)   Add 1 to the number of red tiles.

7 $a = 62°$, $b = 30°$, $c = 30°$, $d = 150°$

8 (a) $12 - (\mathbf{6} + 4) = 2$
   (b) $24 ÷ (2 × \mathbf{4}) = 3$
   (c) $18 - (7 - \mathbf{2}) = 13$

9 (a) T   (b) Q   (c) S   (d) R   (e) P

10 6

11 $w = 2r + 2$ or equivalent

12 (a) 578 m^2 (b) 128 m^2

13 $a = 143°$, $b = 91°$, $c = 89°$, $d = 123°$, $e = 53°$

14 $w = 3r + 2$ (w = number of white tiles, r = number of red tiles)

15 $a = 18.5$ m, $b = 15.5$ m

16 (a) $u = 4$ (b) $v = 3$ (c) $w = 2.5$

17 $x = 14°$

Negative numbers

Essential	**Optional**
Sheets 90 to 92	Large cards with numbers from ⁻9 to 9
Practice sheets P86 to P89	

Ⓐ Colder and colder (p 160)   ▼■●

> Sheets 90 to 92

T

'We had quite a heated (no pun intended) discussion on these and they were amazingly accurate in their final list.'

◊ You could start by asking pupils for examples of temperatures which they think they know. Can they give a reasonable estimate of the room temperature? Have they seen different types of thermometer? Where?

◊ You could ask pupils to mark their suggestions for the temperatures A, B, C, … on a number line. Some pupils may have little idea of some of them.

Temperature trumps (p 163)

This can be played in groups of 2, 3 or 4.

All the cards are dealt, face down. Players do not look at them.

The player whose turn it is reveals their top card and chooses either summer temperature or winter temperature.

The others turn over their top card.

If summer temperature was chosen then the highest summer temperature wins. If winter temperature was chosen then the lowest winter temperature wins.

'Superb game! Worked well with all abilities.'

The winner takes the turned-over cards and has the next turn.

The overall winner is the player with most cards at the end.

B Temperature graphs (p 164)

C Graph or table? (p 165)

D Changes (p 166)

◊ Ask pupils how they see, for example, 2 – 5 on the number line. Help them to associate additions and subtractions with moves on the number line: 2 – 5 as 'start at 2, go down 5'.

◊ It is a good idea to distinguish between the negative sign ⁻ and the subtract sign –, at least to start with.

E A mixed bag of negative numbers (p 169)

E5 There are 336 possible solutions, so do not ask for them all!
When pupils have found a way of completing the first calculation, then the same digits can be fitted into the other two.

F The talent contest (p 170)

This introduces adding and subtracting a negative number.

> Optional: Large cards with numbers from ⁻9 to 9

'This section went well as we did it practically and had joke-telling competitions and all scores were combinations of negative and positives.'

◊ The work is more memorable if you have a real contest!

◊ Focus first on adding together a set of numbers which includes a negative number (which pulls the total score down). Check that pupils realise that, for example, ⁻3 + 5 is equal to 5 + ⁻3.

◊ When addition is understood, get pupils to think about what happens to a total score when a negative number is removed (subtracted), as in the lower pictures on p 171.

Ⓐ **Colder and colder** (p 160)

The approximate temperatures in order are

J   Oven temperature for baking a cake 200°C

I   Temperature of a hot bowl of soup 55°C

G   Temperature of a hot bath 40°C

B   Human body temperature 37°C

H   Temperature at which butter melts 35°C

C   Temperature of a hot summer's day in Britain 30°C

L   Temperature of a heated swimming pool 25°C

E   Temperature inside an ordinary fridge 5°C

K   Temperature of ice-cream when it's good to eat 2°C

F   Temperature inside a car in the morning after a frosty night 1°C

D   Antarctic sea water temperature ⁻1°C

A   A winter's day temperature at the north pole ⁻30°C

A1
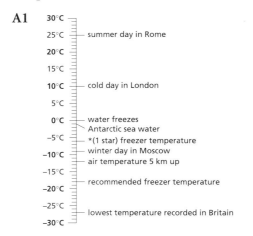

A2 (a) 15°C (b) 25°C (c) ⁻2°C

A3 (a) ⁻8°C (b) ⁻2°C (c) 2°C
    (d) −9°C

A4 (a) 7°C (b) ⁻1°C (c) ⁻5°C
    (d) −50°C

A5 Twelve true sentences are possible.

15°C is	5 degrees higher than	10°C
	10	5°C
	15	0°C
10°C is	5 degrees higher than	5°C
	10	0°C
	15	⁻5°C
5°C is	5 degrees higher than	0°C
	10	⁻5°C
	15	⁻10°C
0°C is	5 degrees higher than	⁻5°C
	10	⁻10°C
⁻5°C is	5 degrees higher than	⁻10°C

A6 Twelve true sentences are possible.

⁻10°C is	5 degrees lower than	⁻5°C
	10	0°C
	15	5°C
⁻5°C is	5 degrees lower than	0°C
	10	5°C
	15	10°C
0°C is	5 degrees lower than	5°C
	10	10°C
	15	15°C
5°C is	5 degrees lower than	10°C
	10	15°C
10°C is	5 degrees lower than	15°C

A7 (a) Scott (b) Nord
    (c) 10 degrees (d) 9 months
    (e) June, July, August
    (f) November, December, January

A8 (a) (i) Helium (ii) Radon
    (b) Argon
    (c) Nitrogen (than argon)
        Helium (than neon)

Ⓑ **Temperature graphs** (p 164)

B1 (a) 7°C (b) 8 p.m.
    (c) ⁻28°C (d) midnight
    (e) 7°C (f) 16

B2 (a) $^-2°C$ (b) $^-13°C$

     (c) 11:30 p.m. and 1 a.m.

     (d) $2\frac{1}{2}$ hours

C Graph or table? (p 165)

C1 (a), (b) Sturge

     (c) Both table and graph are equally easy to use.

C2 Köge (Both table and graph are equally easy to use.)

C3 5 degrees (The table is probably easier to use.)

C4 June, July, August (The graph is probably easier to use.)

C5 November, December, January (The table is easier to use.)

D Changes (p 166)

D1 (a) $^-2 + 3 = \mathbf{1}$ (b) $^-6 + 2 = \mathbf{^-4}$

     (c) $^-8 + 8 = \mathbf{0}$ (d) $^-8 + 10 = \mathbf{2}$

     (e) $^-4 + 7 = \mathbf{3}$ (f) $^-1 + 6 = \mathbf{5}$

D2 (a) $3 - 2 = \mathbf{1}$ (b) $3 - 4 = \mathbf{^-1}$

     (c) $^-5 - 4 = \mathbf{^-9}$ (d) $^-3 - 3 = \mathbf{^-6}$

     (e) $0 - 4 = \mathbf{^-4}$ (f) $4 - 7 = \mathbf{^-3}$

     (g) $^-7 - 4 = \mathbf{^-11}$ (h) $^-3 - 0 = \mathbf{^-3}$

     (i) $^-5 - 2 = \mathbf{^-7}$

D3 (a) $^-3 + 2 = \mathbf{^-1}$ (b) $^-2 - 8 = \mathbf{^-10}$

     (c) $^-5 + 4 = \mathbf{^-1}$ (d) $^-6 - 2 = \mathbf{^-8}$

     (e) $3 - 7 = \mathbf{^-4}$ (f) $^-7 + 3 = \mathbf{^-4}$

D4 (a) $2 - 6 = \mathbf{^-4}$ (b) $1 - 4 = \mathbf{^-3}$

     (c) $^-2 + 3 = \mathbf{1}$ (d) $0 - 3 = \mathbf{^-3}$

D5 (a) $2 - \mathbf{3} = \mathbf{^-1}$ (b) $2 - \mathbf{4} = \mathbf{^-2}$

     (c) $2 - \mathbf{5} = \mathbf{^-3}$ (d) $\mathbf{2} - 7 = \mathbf{^-5}$

     (e) $^-4 - \mathbf{1} = \mathbf{^-5}$ (f) $\mathbf{7} - 5 = 2$

     (g) $\mathbf{8} - 10 = \mathbf{^-2}$ (h) $\mathbf{8} - 4 = 4$

D6 (a) $^-\mathbf{8} + 3 = \mathbf{^-5}$ (b) $^-\mathbf{2} + 4 = 2$

     (c) $4 + \mathbf{3} = 7$ (d) $4 - \mathbf{11} = \mathbf{^-7}$

D7 (a) $50 - 60 = \mathbf{^-10}$ (b) $^-30 - 50 = \mathbf{^-80}$

     (c) $^-\mathbf{40} - 60 = \mathbf{^-100}$ (d) $^-90 + \mathbf{80} = \mathbf{^-10}$

     (e) $15 - \mathbf{115} = \mathbf{^-100}$ (f) $100 - \mathbf{121} = \mathbf{^-21}$

     (g) $\mathbf{60} - 18 = 42$ (h) $^-63 + \mathbf{48} = \mathbf{^-15}$

D8 (a) $10, 8, 6, 4, \mathbf{2}, \mathbf{0}, \mathbf{^-2}$ (subtract 2)

     (b) $7, 5, 3, 1, \mathbf{^-1}, \mathbf{^-3}, \mathbf{^-5}$ (subtract 2)

     (c) $40, 30, 20, 10, \mathbf{0}, \mathbf{^-10}, \mathbf{^-20}$ (subtract 10)

     (d) $^-2, ^-4, ^-6, ^-8, \mathbf{^-10}, \mathbf{^-12}, \mathbf{^-14}$ (subtract 2)

     (e) $^-12, ^-9, ^-6, ^-3, \mathbf{0}, \mathbf{3}, \mathbf{6}$ (add 3)

     (f) The pupil's sequences

D9 Five calculations can be made.

     $^-4 + \mathbf{6} = 2$ $^-4 + \mathbf{5} = 1$ $^-4 + \mathbf{3} = \mathbf{^-1}$

     $^-4 + \mathbf{2} = \mathbf{^-2}$ $^-4 + \mathbf{1} = \mathbf{^-3}$

D10 The four calculations which are most likely to be made are

     $3 - 6 = \mathbf{^-3}$ $3 - 5 = \mathbf{^-2}$ $3 - 2 = 1$

     $3 - 1 = 2$

     These are unlikely to be found at this stage: $3 - ^-2 = 5$, $3 - ^-3 = 6$

D11 (a) 5   (b) $^-1$ (c) 6   (d) $^-5$

     (e) $^-7$ (f) $^-10$ (g) $^-1$ (h) $^-10$

D12 (a) $15 - 20$ (b) $^-5°C$

D13 (a) $^-20 + 38 = 18$ (b) $40 - 42 = \mathbf{^-2}$

     (c) $^-60 + 35 = \mathbf{^-25}$ (d) $20 - 290 = \mathbf{^-270}$

E A mixed bag of negative numbers (p 169)

E1 56 degrees

E2 120°C

E3 34 degrees

E4 (a) (i) 12 degrees   (ii) 12 degrees

       (iii) 6 degrees

     (b) (i) $^-10°C$ (ii) $^-34°C$ (iii) $14°C$

E5 There are 336 solutions, so don't ask for all of them!

In each blank calculation, any triple of digits can be used in which one digit is the sum of the other two.

𝔽 The talent contest (p 170)

F1 (a) ⁻1 (b) ⁻11 (c) 0 (d) ⁻1

F2 (a) 2 (b) ⁻3 (c) 5 (d) 5
(e) ⁻5 (f) 0 (g) 2 (h) 4
(i) 20

F3 (a) 4 (b) 5 (c) 4 − ⁻1 = 5

F4 (a) 3 − ⁻6 = 9 (b) 3 − ⁻3 = 6

F5 (a) Both have the same effect.
(b) Add 2 to 5.
(c) Add the corresponding positive number.
(d) The pupil's examples

F6 (a) 10 (b) 22 (c) 80
(d) 402 (e) 4.2 (f) 8.22

F7 (a) 5 (b) 17 (c) ⁻21
(d) ⁻38 (e) 206 (f) ⁻1.8

F8 (a) 10 (b) ⁻5 (c) 30
(d) 7.7

F9 (a) 1 (b) 11 (c) ⁻2
(d) 78

F10 The possibilities are too numerous to list them all.

F11 Many magic squares are possible.
The 'magic total' is ⁻3.

F12 The 'magic total' is ⁻2.

What progress have you made? (p 173)

1 ⁻3°C

2 ⁻60°C ⁻18°C ⁻6°C 0°C 18°C

3 (a) 5 (b) ⁻2 (c) 0

4 (a) ⁻10°C (b) 22°C

5 (a) Midnight and 8 a.m.
(b) 2 p.m.
(c) 8 p.m., about 9:20 a.m. and 6 p.m.

6 (a) ⁻3 (b) ⁻3 (c) ⁻14 (d) ⁻80

7 (a) **2** − 5 = ⁻3 (b) 10 − **15** = ⁻5
(c) ⁻**6** − 1 = ⁻7

8 (a) 10 (b) 18 (c) ⁻5 (d) 5

Practice sheets

Sheet P86 (section A) ▼□○

1 A, D, C, B, E

2

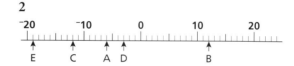

3 (a) ⁻12°C (b) ⁻4°C

4 (a) 0°C (b) 1°C

5 (a) 8°C (b) ⁻8°C

6 (a) ⁻9°C (b) 3°C

Sheet P87 (section B) ▽■●

1 (a) ⁻9°C (b) ⁻6°C
(c) 8 a.m. (and also at roughly 5:20 a.m.)
(d) Noon, 7°C (e) 7 a.m., ⁻13°C
(f) 4 hours (g) 5 hours
(h) 13 degrees (i) 7 degrees

Sheet P88 (section C) ▽■●

1 (a) ⁻3 + 4 = 1 (b) 1 − 4 = ⁻3
(c) ⁻4 + 6 = 2

2 (a) 3 − 6 = **⁻3** (b) ⁻2 − 8 = **⁻10**
(c) 0 − 7 = **⁻7** (d) 2 − 4 = **⁻2**

(e) $3 - \mathbf{5} = {}^{-}2$ (f) $5 - \mathbf{6} = {}^{-}1$

(g) $\mathbf{^{-}7} - 3 = {}^{-}10$ (h) $\mathbf{1} - 2 = {}^{-}1$

3 (a) $^{-}2 + \mathbf{10} = 8$ (b) $^{-}\mathbf{15} + 5 = {}^{-}10$

(c) $^{-}\mathbf{1} + 1 = 0$ (d) $^{-}3 + \mathbf{15} = 12$

4 (a) $100 - 110 = \mathbf{^{-}10}$

(b) $110 - 100 = \mathbf{10}$

(c) $100 - \mathbf{300} = {}^{-}200$

(d) $0 - \mathbf{10} = {}^{-}10$

(e) $25 - \mathbf{100} = {}^{-}75$

(f) $150 - \mathbf{250} = {}^{-}100$

(g) $^{-}20 - \mathbf{0} = {}^{-}20$

(h) $^{-}\mathbf{1} - 8 = {}^{-}9$

(i) $^{-}\mathbf{13} + 7 = {}^{-}6$

5 (a) $180 - 210 = {}^{-}30$

(b) $22 - 150 = {}^{-}128$

(c) $^{-}101 + 67 = {}^{-}34$

Sheet P89 (section E)　　　▽■●

1 (a) 2 (b) 3 (c) 1 (d) $^{-}1$

2 (a) 1 (b) 0 (c) $^{-}7$ (d) 0

(e) $^{-}1$ (f) 3

3 (a) $^{-}1 + {}^{-}\mathbf{3} = {}^{-}4$ (b) $6 + {}^{-}\mathbf{2} = 4$

(c) $4 + {}^{-}\mathbf{3} + {}^{-}1 = 0$ (d) $^{-}\mathbf{4} + 4 = 0$

(e) $5 + {}^{-}\mathbf{6} + 2 = 1$ (f) $^{-}1 + {}^{-}1 + \mathbf{0} = {}^{-}2$

4 (a) 1 (b) $^{-}4$ (c) $^{-}4$

(d) $^{-}2$ (e) 1 (f) $^{-}11$

5 (a) 8 (b) 12 (c) $8 - {}^{-}4 = 12$

6 (a) 14 (b) $10 - {}^{-}4 = 14$

7 (a) $12 - {}^{-}3 = 15$ (b) $6 - {}^{-}5 = 11$

(c) $^{-}1 - {}^{-}6 = 5$

8 (a) 15 (b) 18 (c) 4

(d) 1 (e) 0 (f) 5

9 (a) $^{-}5$ (b) 1 (c) $^{-}3$

(d) 3 (e) 3 (f) $^{-}1$

(g) $^{-}3$ (h) $^{-}2$ (i) $^{-}14$

(j) $^{-}8$ (k) 135 (l) $^{-}57$

10 (a) $^{-}5 - {}^{-}\mathbf{17} = 12$

(b) $2 - {}^{-}\mathbf{8} = 10$

(c) $\mathbf{8} - {}^{-}12 = 20$

(d) $^{-}\mathbf{16} - {}^{-}4 = {}^{-}12$

(e) $^{-}2 + {}^{-}\mathbf{10} = {}^{-}12$

(f) $^{-}2 - \mathbf{10} = {}^{-}12$

(g) $0 + {}^{-}\mathbf{89} = {}^{-}89$

(h) $^{-}15 - {}^{-}\mathbf{30} = 15$

(i) $^{-}\mathbf{5} + {}^{-}45 = {}^{-}50$

(j) $^{-}70 - {}^{-}\mathbf{170} = 100$

(k) $^{-}\mathbf{68} - {}^{-}23 = {}^{-}45$

(l) $\mathbf{0} - {}^{-}45 = 45$

Spot the rule

The teacher-led, whole-class activity that starts off section A is the most important activity in this unit.

At first sight, it may seem perverse not to use input numbers in order, increasing by 1. But this encourages just one method for solving these rule puzzles. An equally effective method may be to ask 'What happens to 100? to 1000?', and pupils may use others.

p 175 **A** Finding rules	▼■●	Finding a rule that connects pairs of numbers	
p 176 **B** Using letters	▼■○	Finding a rule from a table Using *n* to stand for *number*	
p 178 **C** More shorthand	▼■●	Using shorthand such as $4n$ and $\frac{n}{2}$	
p 180 **D** Further rules	▽□●	Using shorthand such as n^2 and $2n^2$	

Practice sheets P90 to P92

Finding rules (p 175)

◊ This section starts with a teacher-led activity with the whole class. This is especially engaging and effective if carried out in complete silence, as the following account from a school makes clear.

'It was the last lesson on Friday afternoon. I allowed time for settling down and said we were going to play a game … everyone had to be silent, including me!

Without saying anything, I wrote on the board …

$$2 \quad \rightarrow \quad 9$$
$$5 \quad \rightarrow \quad 12$$
$$3 \quad \rightarrow$$

… and (still without saying anything) invited someone to come and fill in the gap.

$$3 \quad \rightarrow \quad 10 \qquad ☺$$

As they were right I drew a smiley face.

I then added more numbers each time on the left-hand side. Different pupils came and wrote up numbers on the right-hand side.

3	→	10	☺		
7	→	15	☹	14	☺
10	→	17	☺		
0	→	7	☺		
25	→				

If the pupil's number was wrong I drew a sad face and signalled another pupil to try. They liked big numbers, so I continued with larger ones.

After a while I wrote up …

number   →

… and received the responses

number   →   add 7 ☹

→   number add on 7   ☺

I was looking for the rule here (which could be in words).

After several numbers had been written on the board and pupils had come up and answered in turn, I wrote up a selection of numbers in the left-hand column which included some easy numbers and some more difficult ones.

Pupils came out and filled in the ones they felt comfortable with.

Decimals and negative numbers were included when appropriate.

2	→	9	
5	→	12	
25	→		
109	→		
4	→	11	☹
964	→		
0.5	→	7.5	☹
⁻3.4	→		

After a few games, instead of *number* I introduced the class to using *n* or another letter. I think it is important to emphasise the use of *n* for *number* when pupils are expressing their rules. Initially $n \to + 3$ is quite common instead of $n \to n + 3$.

Once I had played three or four games with the class I asked the pupils to think up some rules of their own. They then took turns to try out their games with the class.

I found there was a beautiful atmosphere with the lesson progressing with no one ever speaking. This can happen quite naturally and seems to strengthen pupils' focus on puzzling out the rule. It allows time for everyone to think and not just the quick ones.'

◊   After playing the game with the whole class, you could continue in groups.

A variation is to set up a simple spreadsheet to produce outputs automatically – nothing complicated, just a formula such as = A1*2 + 1 hidden in cell B1. Pupils can then just type numbers into A1 and see each corresponding result in B1. You could use graphic calculators in a similar way.

B Using letters (p 176) ▼■○

C More shorthand (p 178) ▽■●

D Further rules (p 180) ▽□●

> 'Pupils found this hard – worthwhile keeping it in, though.'

A Finding rules (p 175)

A1 (a) 31 (b) 10 (c) 22 (d) 34
(e) 1 (f) 301

A2 The pupil's results for
$number \rightarrow (number \times 2) - 1$

A3 The pupil's rule and results

B Using letters (p 176)

B1 (a)

$n \rightarrow n - 4$	
7 →	**3**
12 →	**8**
20 →	**16**
4 →	**0**
13 →	9
24 →	20

(b)

$n \rightarrow n \times 5$	
3 →	**15**
8 →	**40**
2 →	**10**
10 →	**50**
4 →	20
6 →	30

(c)

$n \rightarrow n \div 2$	
10 →	**5**
16 →	**8**
22 →	**11**
0 →	**0**
8 →	4
12 →	6

B2 (a) $n \rightarrow (n \times 3) - 1$
(b) 29 (c) 11 (d) 2 (e) 32

B3 (a) $n \rightarrow (n - 2) \times 2$
(b) 6 (c) 8 (d) 20 (e) 0

B4 (a) It could be any of the rules.
(b) $n \rightarrow (n \times 3) - 4$

B5 (a) $n \rightarrow (n \times 4) - 8$
(b) The pupil's numbers following the rule

B6 The pupil's different rules for 3 → 6

B7 (a) $number \rightarrow$ half of $number$
$n \rightarrow n \div 2$

(b)

8 →	**4**
4 →	**2**
12 →	**6**
3 →	$1\frac{1}{2}$ or **1.5**
2 →	**1**
1 →	$\frac{1}{2}$ or **0.5**

B8 (a)

$n \to n \div 3$	
12	**4**
21	**7**
30	**10**
3	**1**

(b)

$n \to n \times 4$	
3	**12**
5	**20**
1	**4**
10	**40**

(c)

$n \to n \div 4$	
8	**2**
12	**3**
20	**5**
2	$\frac{1}{2}$ or **0.5**

B9 (a) 3   (b) 6   (c) 1

(d) 10   (e) $2\frac{1}{2}$ or 2.5   (f) $\frac{1}{2}$ or 0.5

B10 (a) $n \to (n-2) \div 3$ or
$number \to (number - 2) \div 3$

(b) 2   (c) 4   (d) 1

(e) 10   (f) 0

ℂ **More shorthand** (p 178)

C1 (a)

$n \to n + 11$	
3	14
10	**21**
30	**41**
0	**11**
56	**67**
32	43

(b)

$n \to n \div 5$	
15	3
10	**2**
20	**4**
0	**0**
50	10

(c)

$n \to 2n + 1$	
3	7
6	**13**
10	**21**
8	**17**
5	11

(d)

$n \to 10 - n$	
3	7
8	2
4	**6**
9	**1**
5	**5**
7	3

C2 The pupil's rule using n

C3 (a) It could be any of the rules.

(b) $n \to 4n - 1$

C4 (a) $n \to 5n + 1$

(b) 21   (c) 101   (d) 126   (e) 1

C5

$n \to 4n - 2$	
3	10
8	**30**
4	**14**
9	**34**
5	18

C6 (a) $n \to 3n + 2$ (b) $n \to 6n - 1$

(c) $n \to 2n + 3$

C7

$n \to \frac{n}{4}$	
20	5
12	**3**
36	**9**
100	**25**
32	8

C8 (a) 2   (b) 5   (c) 2.5

(d) 0.5   (e) 0

C9 (a) $n \to 5n$ (b) $n \to 3n$

(c) $n \to 2n + 4$ (d) $n \to 6n + 2$

(e) $n \to 4n - 1$ (f) $n \to n + 12$

(g) $n \to \frac{n}{2}$ (h) $n \to 30 - 2n$

(i) $n \to \frac{n}{6}$ (j) $n \to 8 - 2n$

(k) $n \to 100 - 4n$ (l) $n \to \frac{n}{2}$

𝔻 **Further rules** (p 180)

D1

$n \to n^2$	
3	9
10	**100**
5	**25**
4	16

D2(a)

$n \rightarrow n^2 + 1$	
10	→ 101
5	→ **26**
4	→ **17**
9	→ 82

(b)

$n \rightarrow \dfrac{n^2}{2}$	
10	→ 50
6	→ **18**
8	→ **32**
5	→ 12.5

(c)

$n \rightarrow 2n^2$	
10	→ 200
4	→ **32**
6	→ **72**
5	→ 50

D3 (a) $n \rightarrow n^2 + 10$

(b) $n \rightarrow \dfrac{n^2}{4}$ or $\left(\dfrac{n}{2}\right)^2$

(c) $n \rightarrow n^2 - 1$

(d) $n \rightarrow n^2 + n$

D4 The pupil's game

What progress have you made? (p 181)

1 (a)

$n \rightarrow n + 4$	
3	→ 7
6	→ **10**
10	→ **14**
4	→ 8

(b)

$n \rightarrow n \times 3$	
2	→ **6**
4	→ 12
10	→ **30**
3	→ 9

2 $number \rightarrow (number \times 2) + 1$ fits C.

$number \rightarrow number + 3$ fits A.

$number \rightarrow (number \times 3) - 1$ fits B.

3 $number \rightarrow (number \times 2) - 1$

or $n \rightarrow 2n - 1$

4 (a) The pupil's four number pairs for
$n \rightarrow n + 7$

(b) The pupil's four number pairs for
$n \rightarrow n \times 10$

5 (a) (i) The pupil's four number pairs
for $n \rightarrow 2n - 1$

(ii) The pupil's four number pairs
for $n \rightarrow 3n + 10$

(b) (i) $n \rightarrow 2n + 1$

(ii) $n \rightarrow 4n - 10$

6 (a) The pupil's five number pairs for
$n \rightarrow n^2 + 20$

(b) The pupil's five number pairs for
$n \rightarrow n^2 - 2$

Practice sheets

Sheet P90 (sections A, B)   ▼□○

1 (a) 30 (b) 40 (c) 70

(d) 110 (e) 140 (f) 0

2 (a) 7 (b) 11 (c) 21

(d) 5 (e) 3 (f) 27

3 (a)

$number \rightarrow number - 4$	
7	→ 3
10	→ **6**
22	→ **18**
4	→ **0**
100	→ **96**
33	→ **29**

(b)

$number \rightarrow number \div 3$	
12	→ **4**
6	→ **2**
18	→ **6**
30	→ **10**
3	→ **1**
0	→ **0**

4 (a)

$n \rightarrow n + 5$	
4	→ **9**
2	→ **7**
10	→ **15**
13	→ **18**
3	→ 8
20	→ 25

(b)

$n \rightarrow n \times 3$	
2	→ **6**
5	→ **15**
7	→ **21**
4	→ **12**
10	→ 30
1	→ 3

5 (a) $n \rightarrow (n \times 2) + 5$

(b) 9 (c) 13 (d) 25

6 (a) $n \rightarrow (n - 2) \times 3$

   (b) $n \rightarrow (n \div 4)$

   (c) $n \rightarrow n + 12$

   (d) $n \rightarrow (n \times 3) + 7$

4 (a) $n \rightarrow (n \times 3) - 3$

   (b) $10 \rightarrow 27, \ 4 \rightarrow 9, \ 1 \rightarrow 0$

Sheet P91 (section C) ▽■●

1 (a)

$n \rightarrow n - 7$		
10	$\rightarrow$	**3**
20	$\rightarrow$	**13**
8	$\rightarrow$	**1**
33	$\rightarrow$	**26**
17	$\rightarrow$	10
7	$\rightarrow$	0

(b)

$n \rightarrow 3n$		
4	$\rightarrow$	12
5	$\rightarrow$	**15**
8	$\rightarrow$	**24**
13	$\rightarrow$	**39**
10	$\rightarrow$	30
7	$\rightarrow$	21

(c)

$n \rightarrow 8 - n$		
3	$\rightarrow$	5
6	$\rightarrow$	**2**
1	$\rightarrow$	7
$\frac{1}{2}$	$\rightarrow$	$7\frac{1}{2}$
4	$\rightarrow$	4
0	$\rightarrow$	8

2 (a) $n \rightarrow \frac{n}{2} - 1$

   (b) 2   (c) 5   (d) 14   (e) $4\frac{1}{2}$

3 (a)

$n \rightarrow 10n - 2$		
3	$\rightarrow$	**28**
4	$\rightarrow$	**38**
7	$\rightarrow$	**68**
8	$\rightarrow$	**78**
10	$\rightarrow$	98

(b)

$n \rightarrow \frac{n}{3}$		
15	$\rightarrow$	**5**
21	$\rightarrow$	**7**
60	$\rightarrow$	**20**
36	$\rightarrow$	**12**
27	$\rightarrow$	9

   (c) *number* $\rightarrow$ 10 multiplied by *number* subtract 2

   (d) *number* $\rightarrow$ *number* divided by 3

4 (a) $n \rightarrow (n \times 2) + 1$ or $n \rightarrow 2n + 1$

   (b) $n \rightarrow (4 \times n) - 1$ or $n \rightarrow 4n - 1$

5 *number* $\rightarrow$ *number* times 3 and $n \rightarrow 3n$
   number $\rightarrow$ (4 times *number*) add 3 and $n \rightarrow 4n + 3$
   number $\rightarrow$ (*number* add 3) times 4 and $n \rightarrow (n + 3) \times 4$

6 (a) $n \rightarrow 7n$ or $n \rightarrow 7 \times n$

   (b) $n \rightarrow \frac{1}{2}n$ or $n \rightarrow n \div 2$ or $n \rightarrow \frac{n}{2}$

   (c) $n \rightarrow (n \times 2) - 3$ or $n \rightarrow 2n - 3$

   (d) $n \rightarrow 2n + 10$ or $n \rightarrow (2 \times n) + 10$

Sheet P92 (section D) ▽□●

1 (a)

$n \rightarrow n^2 + 2$		
3	$\rightarrow$	11
5	$\rightarrow$	**27**
8	$\rightarrow$	**66**
10	$\rightarrow$	**102**
2	$\rightarrow$	**6**
12	$\rightarrow$	**146**

(b)

$n \rightarrow n^2 + n$		
1	$\rightarrow$	2
5	$\rightarrow$	**30**
10	$\rightarrow$	**110**
6	$\rightarrow$	**42**
9	$\rightarrow$	**90**
4	$\rightarrow$	**20**

(c)

$n \rightarrow n \times (n + 1)$		
1	$\rightarrow$	**2**
5	$\rightarrow$	**30**
10	$\rightarrow$	**110**
6	$\rightarrow$	**42**
9	$\rightarrow$	**90**
4	$\rightarrow$	**20**

2 (a) $n \rightarrow 3n^2$ and $n \rightarrow n^2 + 2$ could be her rules.

   (b) The pupil's three rules

   (c) $n \rightarrow 3n^2$ is being used.
     $2 \rightarrow$ **12**, $10 \rightarrow$ **300**, $7 \rightarrow$ **147**

Gravestones

Essential

Sheets 93 (▽■●), (94 ▼□○), 95 (▽■●)

Practice sheets P93 to P95

A What gravestones tell us (p 182) ▼■●

'*A really good introduction – it captivated pupils' imagination.*'

◊ In some circumstances, such as a recent bereavement, sensitive handling is needed.

The topic obviously lends itself to locally based practical work. The sample you get from a graveyard is only of people rich enough to afford a gravestone.

A3 There may be discussion about which months are in winter.

B Making a frequency table (p 183) ▼■●

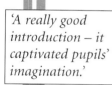

◊ Ask pupils to think of two different ways of tallying from a list.

- Go through the list, tallying all those in the 0–9 group first; then do the 10–19 group, and so on.
- Go through the list once only, tallying into the correct group as you go.

Items are less likely to be missed out if the second way is used.

◊ Pupils may already be familiar with grouping tally marks in fives: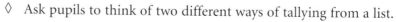

C Comparing charts (p 184)

◊ This can be done in pairs or as a class discussion.

◊ Two ways of drawing a bar chart are shown. In the first, age is treated as discrete and a gap is left between bars. In the second age is treated as continuous with no gaps between bars, so that a rule is needed for the boundaries.

D Charts galore! (p 186)

Pupils look at good and bad ways of displaying data.

◊ This section will mean more if pupils can use a spreadsheet and draw charts for themselves. In one school, the data on sheet 95 for 'Old Faithful' (question E4) was put into Microsoft Excel. Pupils were asked to do an appropriate graph and an inappropriate graph and to explain their choices.

◊ The tilted pie chart can make if difficult to judge the relative sizes of sectors. It loses the empty age groups. It gives no idea of the total sample, but shows the proportions.

The cobweb diagram is hard to read. The shading has no meaning.

The bar chart clearly summarises the data but there is no point in dividing the frequency scale into fifths.

The line graph or frequency polygon retains all the information but can be confusing to interpret. The joining lines have no meaning; they just guide the eye.

The doughnut diagram loses the 10–19 and 20–29 data but shows a gap labelled 30–39!

The 3-D bar chart retains all the information, but the frequency is hard to read.

E Testing a hypothesis (p 188)

Sheet 93

The idea of a hypothesis is difficult. Low attainers can leave this out and go to section F.

Challenge (p 188)

John Millington died on 1 September 1694 aged 54.

If he died before his birthday that year then he was born in 1639. If he died after his birthday that year then he was born in 1640.

F Every chart tells a story (p 189)

This work consolidates tallying and drawing frequency graphs.

> Sheets 94 (▼□○) and 95(▽■●)

It is better if pupils work in pairs, with one reading and the other tallying.

F2 In one class a pupil got fed up with being pestered for his favourite number. He wrote it on a card with 'Do not disturb'!

A What gravestones tell us (p 182)

A1 Probably November 1757. Remember that people are not generally buried on the day they die.

A2 Roughly 90 years

A3 It depends on what are the 'winter months'.

J	F	M	A	M	J	J	A	S	O	N	D
1		2	1		2	2	1		1	3	1

B Making a frequency table (p 183)

B1

Age (in years)	Tally	Frequency
0–9	‖‖ ‖	6
10–19		0
20–29		0
30–39		0
40–49	‖	1
50–59	‖‖‖	3
60–69	‖‖‖	3
70–79	‖	1

The table shows that the people tended to die young or old but not in between.

C Comparing charts (p 184)

C1 There are no missing bars. They have zero height.

C2 70–80

C3 There are no people who died in these age groups.

C4 (a) Pie chart
     (b) Either bar chart
     (c) Either bar chart

E Testing a hypothesis (p 188)

E1 Some pupils will not manage to make a hypothesis of their own. In E2 they can test one of those already suggested.

Possible hypotheses include

'Nobody lived beyond 71.'

'Nobody died aged between 10 and 39.'

'Most people died in November.'

'Boys stood a better chance of surviving childhood.'

E2 'Most deaths occurred in the winter months.'

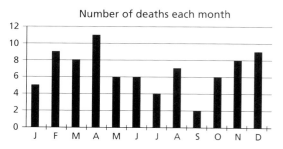

Number of deaths each month

If we define the winter months to be from November to April then the hypothesis seems to be confirmed with 50 dying in that period, 31 outside it.

'Most deaths occurred in the under 10 age group.'

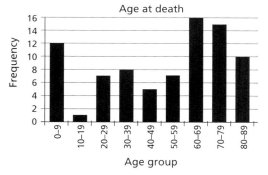

This hypothesis is not confirmed. The bar chart shows that most deaths occurred in the 60–69 age group. It is worth commenting that infant mortality was high.

'If you reached 20, there was a good chance of living to 60.'

Of the 68 people who reached age 20, 41 reached age 60. So the hypothesis is confirmed.

F Every chart tells a story (p 189)

F1

Age group	Tally	Frequency
0–19	||	2
20–39	|||| |	6
40–59	|||| |||	8
60–79	|||| |||| |||| ||	17
80–99	|||	3

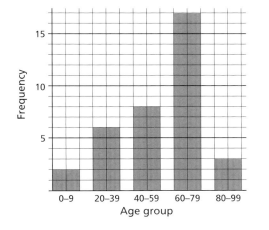

F2 (a) 3 and 7

(b) 33 people

(c) People tend to pick 3 and 7.

(d) The pupil's plan to test their hypothesis

F3 (a)

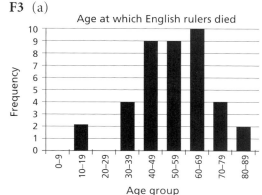

(b) The chart shows that the rulers did not die particularly young. (This could be biased because you might have had to wait for an old parent to die before becoming ruler.) Most managed to get into their forties. (It would be interesting to compare this with the general population.)

F4 (a)

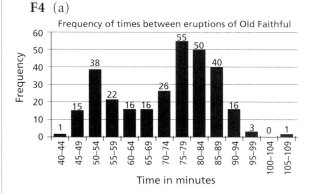

The appearance of the chart will depend on the choice of interval for grouping. This is a point worth discussing.

(b) The chart above shows two peaks to the activity: a minor one after about 50 minutes, and a major one after about $1\frac{1}{4}$ hours. The times between eruptions are in the range 40 to 110 minutes.

(c) You will have to wait at least 40 minutes, but no longer than 110 minutes. You may be lucky and see one after about 55 minutes, but most occur around 75 minutes after the last eruption.

What progress have you made?
(p 190)

1 (a) 12

    (b) 48

    (c) July and August

2

Estimate	Tally	Frequency
5–9	|||	3
10–14	ⵉ⵬ ⵉ⵬ ⵉ⵬		16
15–19	ⵉ⵬		6
20–24	||||	4
25–29		0
30–34			1

Practice sheets

Sheet P93 (section F) ▼□○

1 (a)

Number of tomatoes on plant	Tally	Frequency
0			1
1				2
2						4
3	ⵉ⵬	5
4	ⵉ⵬ ⵉ⵬	10
5	ⵉ⵬		6
6			1
7			1

(b)

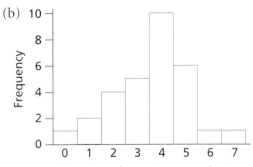

2 (a) 50

    (b) 20

    (c) The greenhouse plants did better. They had more plants with larger numbers of tomatoes on.

Sheet P94 (section F) ▽■○

1 (a) 8   (b) 6   (c) 82

2 (a)

Age at death (years)	Tally	Frequency
0–9	ⵉ⵬ ||||	9
10–19	ⵉ⵬ ⵉ⵬		11
20–29	ⵉ⵬ ⵉ⵬ ⵉ⵬ ⵉ⵬		21
30–39	ⵉ⵬ |||	8
40–49	|||	3
50–59	||||	4
60–69			1
70–79			1
80–89		0
90–99			1

(b)

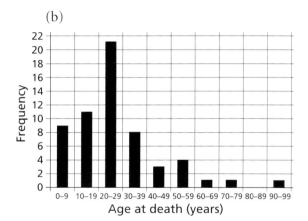

(c) The pupil's two sentences, such as:

About two-thirds of the females died before 30 years of age.

Male deaths were more evenly spread.

About half the males died before 30 years of age.

A much larger proportion of the males lived beyond 60.

Sheet P95 (section F)   ▽ □ ●

1 Pupils may choose groups of a different size from those shown here.

With fertiliser

Weight of turnips (grams)	Tally	Frequency
100–149	ЖII	6
150–199	Ж Ж II	12
200–249	Ж II	7
250–299	IIII	4
300–349	I	1

Without fertiliser

Weight of turnips (grams)	Tally	Frequency
100–149	II	2
150–199	IIII	4
200–249	Ж I	6
250–299	Ж Ж	10
300–349	Ж III	8

2

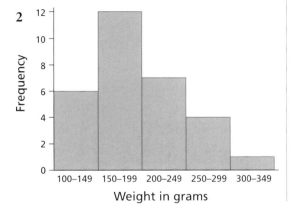

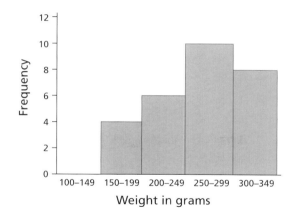

3 The fertiliser does give bigger turnips.

Over half the turnips grown without fertiliser are under 200 grams whereas when fertiliser has been used about two-thirds are over 250 grams.

Number patterns

Essential

Dice (▼□○)

Optional

Sheet 96
Multilink or other cubes

Practice sheets P96 to P101

A Exploring a number grid (p 191)

These investigations are all based on the six-column grid and are graded in difficulty. Some teachers have preferred to do some or all of them later in the unit.

> Optional: Sheet 96

Investigation 1 (Add a number in column A to one in column B)

◊ This is a good one to start on with the whole class. It leads on to variations which pupils can investigate for themselves.

The result of A + B is always in column C (provided the grid is extended downwards). You can then ask pupils what they think could be meant by 'investigate further'. All suggestions should be responded to positively, even though they may lead nowhere.

Here are some fruitful suggestions:

What if we add two different columns?
What if we subtract?
What if we multiply?

Investigation 2 (Multiples)

◊ These occur in spatially regular patterns.

Investigation 3 (Predict the 30th number in column B, etc.)

◊ There are various ways to do this. One is to work out the 30th number in column F ($30 \times 6 = 180$) and then work back to 176.

Investigation 4 (Predict which column 500 will be in, etc.)

◊ $500 \div 6 = 83$ remainder 2, so 500 will be in column 2.

Investigation 5 (Prime numbers)

◊ Prime numbers (except for 2 and 3) are only in columns A and E. This is because the numbers in the other columns are either even or multiples of 3.

Ⓑ **Dice numbers** (p 192)   ▼□○

> Dice (one per pupil)

Ⓒ **Magic squares** (p 193)   ▼□○

Ⓓ **Rectangles** (p 194)   ▼■○

Prime numbers are introduced as numbers which cannot be made into a rectangular array, only a single line.

Ⓔ **Square numbers** (p 195)   ▼■●

Ⓕ **Cubes** (p 196)   ▽■●

> Optional: Multilink or other types of cube

Ⓖ Square roots (p 197)

Ⓗ Missing numbers (p 199)

All the sequences in this section are linear.

◊ To vary the way in which tasks are presented, you could think up a sequence and write each number (for example, 6, 10, 14, 18, 22, 26, 30, 34) on a card or piece of paper without showing the pupils. Shuffle the cards, remove two and then show the rest (or read out the numbers) in no particular order. The pupils have to decide what numbers are on the missing cards. (Sometimes, if the highest or lowest card is missing, they could be left with a choice of missing numbers both of which would fit.)

Pupils could then make up their own sequence cards and do the activity with each other.

H7 This is harder. Pupils could use trial and improvement.

Ⓘ Sequences (p 200)

Ⓙ Sequences in tables (p 201)

Ⓚ Even number, odd number (p 202)

Ⓑ Dice numbers (p 192)

B1 (a) 2   (b) 4

    (c) Top and bottom numbers add up to 7.

B2 (a) Can see 7, can't see 14

    (b) Can see 12, can't see 9

    (c) Can see 10, can't see 11

    (d) Can see 6, can't see 15

B3 (a) Yes   (b) 6, 7, 9, 10, 12, 14, 15

    (c) 15   (d) No

B4 (a) 27   (b) 26   (c) 26   (d) 25

Ⓒ Magic squares (p 193)

C1 15

C2 (a)

7	2	9
8	6	4
3	10	5

(b)

6	11	4
5	7	9
10	3	8

(c)

10	5	6
3	7	11
8	9	4

C3 (a)

7	14	**9**
12	**10**	8
11	**6**	**13**

(b) The new square is a magic square. The magic number is 30 because there is an extra 15 in each row, etc.

C4 (a)

16	3	2	13
5	**10**	**11**	8
9	**6**	7	**12**
4	15	**14**	1

(b)

15	**10**	3	**6**
4	5	**16**	9
14	11	2	**7**
1	**8**	13	12

D **Rectangles** (p 194)

D1 3×8 (or 8×3). Don't count a single line (1×24) as a rectangle.

D2 10×2

D3 (a) 2×6, 3×4
(b) 2×8, 4×4
(c) 2×9, 3×6
(d) 2×15, 3×10, 5×6

D4 You can only make a single line with 17. Other numbers include 2, 3, 5, 7, …

D5 11, 13, 17

D6 Because even numbers can be made into a rectangle $2 \times$ something

D7 (a) 11 (b) 17 (c) 19 (d) 23
(e) 29 (f) 31 (g) 37

E **Square numbers** (p 195)

E1 $6 \times 6 = 36$ and $7 \times 7 = 49$

E2 64, 81, 100

E3 3, 5, 7, 9, … odd numbers

E4 (a) 16 (b) 25 (c) 121 (d) 400

E5 Rob has done 10×2. He should have done 10×10.

E6 (a) 13 (b) 33 (c) 73
(d) 61 (e) 155

E7 $6 = 2^2 + 1^2 + 1^2$
$7 = 2^2 + 1^2 + 1^2 + 1^2$
$8 = 2^2 + 2^2$
$9 = 3^2$
$10 = 3^2 + 1^2$
$11 = 3^2 + 1^2 + 1^2$
$12 = 2^2 + 2^2 + 2^2$
$13 = 3^2 + 2^2$
$14 = 3^2 + 2^2 + 1^2$
$15 = 3^2 + 2^2 + 1^2 + 1^2$
$16 = 4^2$
$17 = 4^2 + 1^2$
$18 = 3^2 + 3^2$
$19 = 3^2 + 3^2 + 1^2$
$20 = 4^2 + 2^2$
$21 = 4^2 + 2^2 + 1^2$
$22 = 3^2 + 3^2 + 2^2$
$23 = 3^2 + 3^2 + 2^2 + 1^2$
$24 = 4^2 + 2^2 + 2^2$
$25 = 5^2$
$26 = 5^2 + 1^2$
$27 = 5^2 + 1^2 + 1^2$ or $3^2 + 3^2 + 3^2$
$28 = 5^2 + 1^2 + 1^2 + 1^2$ or
$\quad\quad 4^2 + 2^2 + 2^2 + 2^2$ or
$\quad\quad 3^2 + 3^2 + 3^2 + 1^2$
$29 = 5^2 + 2^2$
$30 = 5^2 + 2^2 + 1^2$

E8 (a) $7^2 = 1 + 3 + 5 + 7 + 9 + 11 + 13$

    (b) 400

    (c) $9^2 = 1 + 2 + 3 + 4 + 5 + 6 + 7 + 8 + 9 + 8 + 7 + 6 + 5 + 4 + 3 + 2 + 1$

       $9^2 = 1 + 8 + 16 + 24 + 32$

F Cubes (p 196)

F1 (a) 27   (b) 3^3

F2 (a) 72   (b) 150   (c) 152   (d) 657

F3 The differences are 7, 19, 37, 61, 91, …
The differences between the differences are 12, 18, 24, 30, …
These are the multiples of 6, starting with 12.

F4 3375 (15^3)

G Square roots (p 197)

G1 (a) 3   (b) 5   (c) 10
    (d) 8   (e) 1

G2 (a) 2   (b) 9   (c) 6

G3 (a) 7   (b) 12   (c) 21
    (d) 123

G4 31

G5 (a) 3000

    (b) The answer depends on how many people can occupy a square metre.
If it is 1, then the length of the side would be 3 km.

G6 On the same basis, the area would be 6400 km^2 (side 80 km).

Trial and improvement working should be shown in questions G7–G9.

G7 $37^2 = 1369$, so $\sqrt{1369} = 37$

G8 (a) 27   (b) 41   (c) 57   (d) 921

G9 $\sqrt{20} = 4.472\,135\,95…$

G10 (a) 300 km

    (b) 629 m = 0.629 km, so you can see 79.3 km

True or false?

1 False; for example, $\sqrt{0.25} = 0.5$

2 False; for example, $0.5^2 = 0.25$

3 False; $2.5^2 = 6.25$, which is not halfway between 4 and 9

H Missing numbers (p 199)

H1 (a) 17, 20   (b) add 3

H2 (a) 36, 43   add 7
    (b) 11, 7   subtract 4
    (c) 56, 67   add 11
    (d) 28, 21   subtract 7

H3 (a) 2, 8, 14, **20**, 26, **32**, 38
    (b) 5, 9, **13**, **17**, 21, **25**, 29
    (c) 2, 11, 20, **29**, 38, 47, **56**
    (d) 36, 31, **26**, **21**, 16, 11, **6**

H4 15

H5 (a) **14**, **17**, 20, 23, **26**, **29**, 32
    (b) **54**, **46**, 38, 30, **22**, 14, **6**

H6 29, 47

H7 (a) 7, **10**, 13, **16**, 19, **22**, 25
    (b) 1, **5**, 9, **13**, **17**, 21, **25**

I Sequences (p 200)

I1 (a) 12   (b) Even numbers, or add 2

I2 (a) Add 4   (b) Subtract 3

I3 (a) 7, **11**, 15, **19**, 23, 27, **31**   add 4
    (b) 1, **4**, 7, **10**, **13**, 16, **19**   add 3
    (c) **41**, **37**, 33, **29**, 25, **21**, 17   subtract 4
    (d) 44, **39**, 34, **29**, 24, 19, **14**   subtract 5

I4 (a) Add 1, then 2, then 3 …
next number 22

(b) Add 2, then 4, then 6, then 8, …
next number 45

(c) Subtract 3, then 6, then 9, then 12, …
next number 22

(d) Add 5, then 7, then 9, then 11, …
next number 62

(e) Subtract 1, then 3, then 5, then 7, …
next number ⁻5

I5 $1 + 2 = 3$, $2 + 3 = 5$, $3 + 5 = 8$,
$5 + 8 = 13$, …

Challenge!

(a) 1, 2, 4, 8, 16, 32, **64**, **128**
double each time

(b) 1, **4**, **9**, 16, **25**, 36, 49, **64**
square numbers

(c) 2, 3, 5, 7, 11, 13, 17, **19**, **23**
prime numbers

♩ Sequences in tables (p 201)

J1 (a) Square numbers

(b) Square numbers + 1

(c) Add 2, then 4, then 6, then 8, …
(add even numbers)

J2 (a) Add 4, then 8, then 12, …
(add multiples of 4);
?, ? are 61, 85.

(b) The pupil's investigation

J3 On every diagonal the differences go up
by 8 each time.

81, 121 (odd square numbers)

65, 101 (even squares + 1)

57, 91

73, 111

𝕂 Even number, odd number (p 202)

The diagrams suggest thinking of an even
number as a set of twos, and an odd
number as a set of twos plus an extra one.

So an odd number plus an odd number is
two sets of twos plus two ones; the two
ones make a two, so you are left with only
twos, giving an even number.

Similar reasoning can be used for the
sum of two consecutive numbers and the
pattern of odd and even in triangle
numbers.

What progress have you made? (p 202)

1 Second row × 2 is in first row.
First row × 2 is in second row.
Third row × 2 is in third row.

2 (a) 5, 9, 13, 17, 21, **25**, **29**   add 4

(b) 43, 37, 31, 25, 19, **13**, **7**   subtract 6

(c) **10**, **17**, 24, 31, 38, 45, **52**   add 7

3 (a) 25 is 5×5 (b) 16   (c) 36

4 (a) 13 can only be divided by 1 and
by 13. It can't be arranged in a
rectangle, only a line.

(b) 23, 29

5 (a) Subtract 6
Next numbers 22, 16

(b) Add 2, 4, 6, 8, (add even numbers)
Next numbers 43, 57

6 64

7 (a) Because $12 \times 12 = 144$

(b) The square of 9 means 9×9.
The square root of 9 is the number
whose square is 9, that is 3.

8 53.09…

9 The pupil's investigation

Practice sheets

Sheet P96 (section B, C)

1 3

2 12

3 (a) 10   (b) 15   (c) 14

4 (a) 23   (b) 32   (c) 24

5

1	14	4	15
8	11	5	10
13	2	16	3
12	7	9	6

Sheet P97 (section D, E)

1 (a) Rectangle pattern 3×6

   (b) Rectangle pattern 2×16

2 (a) 2×4

   (b) 2×14, 4×7

   (c) 2×6, 3×4

   (d) 2×18, 3×12, 4×9, 6×6

3 No, 9 makes a square 3×3.

4 (a) 23   (b) 29   (c) 7

5 (a) 29, 31   (b) 41   (c) 47

6 (a) 9 (b) 49 (c) 81 (d) 100

7 5^2 obviously makes a square. Prime numbers don't make rectangles (or squares).

8 (a) 7 (b) 16 (c) 37 (d) 36

9 16, 1, 9 and 25

Sheet P98 (sections D, E, F)

1 (a) 2×18, 3×12, 4×9, 6×6

   (b) 2×20, 4×10, 5×8

   (c) 2×11

   (d) 2×25, 5×10

2 (a) 29   (b) 7   (c) 41

3 (a) 47   (b) 59, 61   (c) 97

4 (a) 9   (b) 49   (c) 81

   (d) 100   (e) 64

5 5^2 obviously makes a square. Prime numbers don't make rectangles (or squares).

6 (a) 7 (b) 16 (c) 37 (d) 36

7 125

8 (a) 27   (b) 343   (c) 1000

   (d) 8

9 44 $(6^2 + 2^3)$, 64 $(10^2 - 6^2)$, 125 (5^3), 128 $(4^3 + 4^3)$, 225 (15^2), 243 $(3^3 + 6^3)$, 271 $(10^3 - 9^3)$

Sheet P99 (sections E, F, G)

1 (a) 9 (b) 49 (c) 81 (d) 10 000

2 (a) 27 (b) 56 (c) 257 (d) 19

3 (a) 27   (b) 1000   (c) 1331

   (d) 125

4 44 $(6^2 + 2^3)$, 64 $(10^2 - 6^2)$, 125 (5^3), 128 $(4^3 + 4^3)$, 225 (15^2), 243 $(3^3 + 6^3)$, 271 $(10^3 - 9^3)$

5 (a) 1936 (44×44)

   (b) 1728 $(12 \times 12 \times 12)$

6 (a) 4 (b) 7 (c) 2 (d) 12

7 4000

8 607

9 (a) 2 seconds

   (b) About 6 seconds (6.32…)

Sheet P100 (section H)

1 (a) 22 and 26

   (b) You add 4 each time.

2 (a) 25, 28   add 3

   (b) 30, 26   take off 4

   (c) 80, 93   add 13

   (d) 45, 34   take off 11

3 (a) 28, 40   (b) 31, 34, 40

  (c) 14, 32, 50   (d) 44, 41, 32

4 17

5 (a) 38, 26   (b) 16, 46

Sheet P101 (section I)

1 (a) 25, 28   add 3

  (b) 30, 26   take off 4

  (c) 80, 93   add 13

  (d) 45, 34   take off 11

2 (a) 28, 40 add 6

  (b) 31, 34, 40 add 3

  (c) 14, 32, 50 add 6

  (d) 47, 44, 32 take off 3

3 (a) 38, 26   (b) 16, 46

4 (a) Next two numbers 33, 45.
    Add 2, then 4, then 6, then 8, …

  (b) Next two numbers 90, 126.
    Add 6, then 12, then 18, then 24, …

  (c) Next two numbers 51, 66.
    Add 3, then 5, then 7, then 9,
    then 11, …

  (d) Next two numbers 58, 44.
    Take off 2, then 4, then 6, then 8,
    then 10, …

Review 4 (p204)

1 (a) The pupil's pattern for 36

(b) 2×18, 3×12, 4×9, 6×6

(c) Yes, $36 = 6^2$

2 (a) 7   (b) 14   (c) 36

3 (a) 8.5 cm, 8.5 cm, 2.9 cm

(b) 20°, 80°, 80°

(c) Isosceles

4 (a) 5°C   (b) 15 degrees   (c) 30 degrees

5 (a) $3 \times (\mathbf{6} - 1) = 15$

(b) $(12 - \mathbf{0}) \times 2 = 24$

(c) $4 - (\mathbf{3} - 1) = 2$

7 (a) 16 cm²   (b) 18 cm

(c)

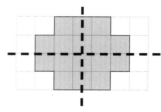

8 (a) The pupil's accurate drawings

(b) AB = 9.5 cm, BC = 3.9 or 4 cm, angle at C = 78°
PR = 8.6 or 8.7 cm, angle at Q = 60°, angle at R = 30°

(c) ABC is isosceles; PQR is scalene and right-angled.

9 (a) 2   (b) ⁻7   (c) ⁻4   (d) ⁻6

(e) ⁻5   (f) 6   (g) ⁻10   (h) ⁻1

10 (a) Add 5   (b) 22, 27, 32

11 (a) Add 4   (b) ⁻5, ⁻1, 3

12 (a) 30 m²   (b) 28 m

13 The pupil's rectangle 4 cm by 12 cm

14 (a) 49   (b) 125   (c) 29

     (d) 97

15 (a) Foot length = 3.3 × index finger length

     (b) 31.5 cm   (c) 9.1 cm

16 $a = 53°$ $b = 73°$ $c = 54°$

     $d = 53°$ $e = 73°$ $f = 54°$

17 (a) ⁻11   (b) 5   (c) ⁻5   (d) ⁻5

     (e) 11   (f) 11   (g) 5   (h) ⁻5

18 (a) $5 - (3 - \mathbf{0}) = 2$

     (b) $5 - (3 - \mathbf{^{-}4}) = {^{-}}2$

     (c) $5 + (3 - \mathbf{10}) = {^{-}}2$

     (d) $5 + (3 - \mathbf{6}) = 2$

     (e) $5 - (3 + \mathbf{0}) = 2$

     (f) $5 - (3 + \mathbf{4}) = {^{-}}2$

19 (a) 4   (b) 12   (c) 36

20 (a) ⁻4, ⁻1, **2**, **5**, **8**

     (b) 2, ⁻**5**, ⁻**12**, ⁻**19**

     (c) 8, 4, 2, **1**, $\frac{1}{2}$, $\frac{1}{4}$

     (d) 1000, 100, 10, **1**, **0.1** or $\frac{1}{10}$, **0.01** or $\frac{1}{100}$

21 The pupil's rectangle 4 cm by 12.5 cm

22 (a) 8.64 cm²   (b) 240 cm²

23 (a) A 1p coin has a diameter of 2 cm.
       So in 1.6 km we can fit 80 000 coins,
       or £800 worth.

     (b) A typical heart would beat about 70 times per minute.
       So a million heart beats would last about 14 000 minutes
       or about 10 days.

24 (a) There is 1 three by three square, 4 two by two squares and 9 one by one squares,
       a total of $1^2 + 2^2 + 3^2 = 14$ squares.

     (b) There is 1 ten by ten, 4 nine by nine, 9 eight by eight, 16 seven by seven,
       25 six by six etc, a total of $1^2 + 2^2 + 3^2 + 4^2 + 5^2 + 6^2 + 7^2 + 8^2 + 9^2 + 10^2 = 385$ squares.

25 There are no unique answers, but …

John may be going up in 2s.

Paul may be adding the previous two numbers together.

Sue may be going 'multiply by 2, add 2' alternately.

26 (a)   9567
+1085
─────
10652

S = 9, E = 5, N = 6, D = 7, M = 1, O = 0, R = 8, Y = 2

(b)   96233
+62513
──────
158746

C = 9, R = 6, O = 2, S = 3, A = 5, D = 1, N = 8, G = 7, E = 4

27 7:10 a.m.

28 (a)   6 ways   (b)   20 ways

29 (a)   Always even. We are multiplying an even number by an odd one.

(b)   One number must be even and one must be divisible by 3.

(c)   24. One must be divisible by 3, one by 2 and one by 4.

30 7, 8, 10, 10 and 11